COMPUTATIONAL FLUID DYNAMICS SIMULATION OF SPRAY DRYERS

AN ENGINEER'S GUIDE

Advances in Drying Science & Technology
Series Editor: Arun S. Mujumdar

PUBLISHED TITLES

Computational Fluid Dynamics Simulation of Spray Dryers:
An Engineer's Guide
Vasile Minea

Advances in Heat Pump-Assisted Drying Technology
Vasile Minea

COMPUTATIONAL FLUID DYNAMICS SIMULATION OF SPRAY DRYERS

AN ENGINEER'S GUIDE

Meng Wai Woo

CRC Press
Taylor & Francis Group
Boca Raton London New York

CRC Press is an imprint of the
Taylor & Francis Group, an **informa** business

CRC Press
Taylor & Francis Group
6000 Broken Sound Parkway NW, Suite 300
Boca Raton, FL 33487-2742

First issued in paperback 2019

© 2017 by Taylor & Francis Group, LLC
CRC Press is an imprint of Taylor & Francis Group, an Informa business

No claim to original U.S. Government works

ISBN-13: 978-1-4987-2464-7 (hbk)
ISBN-13: 978-0-367-87317-2 (pbk)

Visit the Taylor & Francis Web site at
http://www.taylorandfrancis.com

and the CRC Press Web site at
http://www.crcpress.com

Contents

Series Preface

Advances in Drying Science and Technology

It is well known that the unit operation of drying is a highly energy-intensive operation encountered in diverse industrial sectors ranging from agricultural processing, ceramics, chemicals, minerals processing, pulp and paper, pharmaceuticals, coal polymer, food, forest products industries as well as waste management. Drying also determines the quality of the final dried products. The need to make drying technologies sustainable and cost effective via application of modern scientific techniques is the goal of academic as well as industrial R&D activities around the world.

Drying is a truly multi- and interdisciplinary area. Over the last four decades the scientific and technical literature on drying has seen exponential growth. The continuously rising interest in this field is also evident from the success of numerous international conferences devoted to drying science and technology.

The establishment of this new series of books titled *Advances in Drying Science and Technology* is designed to provide authoritative and critical reviews and monographs focusing on current developments as well as future needs. It is expected that books in this series will be valuable to academic researchers as well as industry personnel involved in any aspect of drying and dewatering.

The series will also encompass themes and topics closely associated with drying operations, for example, mechanical dewatering, energy savings in drying, environmental aspects, life cycle analysis, technoeconomics of drying, electrotechnologies, control and safety aspects, and so on.

Dr. Arun S. Mujumdar is an internationally acclaimed expert in drying science and technologies. He is the founding chair (1978) of the International Drying Symposium (IDS) series and has been editor-in-chief of *Drying Technology: An International Journal* since 1988. The 4th enhanced edition of his *Handbook of Industrial Drying* published by CRC Press has just published. He is the recipient of numerous international awards including honorary doctorates from Lodz Technical University, Poland and University of Lyon, France.

Please visit www.arunmujumdar.com for further details.

Dr. Arun S. Mujumdar

Preface

I thank my PhD supervisor Professor Wan Ramli Wan Daud at the University Kebangsaan Malaysia for introducing me to CFD modeling of spray dryers in 2006. The seed to write a book related to CFD was planted some time in 2007, when I began to write my own notes to help me understand the CFD technique. At that time, I was not sure what type of CFD book would emerge, but I was particularly intrigued by the elegance of numerical solutions. I was largely inspired by a seminal book that I stumbled upon, on numerical solution by Patankar (1980).

As I progressed through my PhD on the modeling of spray dryers, I discovered many more numerical intricacies that transform a general CFD simulation into a spray dryer CFD simulation. A large part of what fascinated me about these intricacies, was the understanding of the physics involved and how they are elegantly implemented in the numerical framework. Therefore, at that time, I thought that it would be useful to share some of these numerical experiences. Putting on my hat as an educator, I pondered on how some of this information could be shared to practicing engineers who have not undertaken more advanced postgraduate studies. The shape of this book then started to emerge.

In 2010, when I joined Monash University, I was very fortunate to be able to interact with the dairy industry on the modeling of spray dryers. Very special thanks to Professor

Xiao Dong Chen and Professor Cordelia Selomulya for these opportunities. Professor Chen was my postdoctoral supervisor in Monash University. Working with the industry made me realize that there is a gap in understanding between the spray-drying industry and the numerical modeler on spray drying. This further cemented my conviction to write this book and a more definite shape of the book emerged; not just to share the numerical experience but to also provide an easy read for industry. On the flip side, I thought that it would also be useful for the numerical modeler to understand the industrial "language" when engaging on a CFD project. These cumulated in Chapter 8 of this book and also augmented the tone of this book.

Hence, this book on CFD modeling of spray dryers was produced. I greatly appreciate Professor Arun Sadhashiv Mujumdar for giving me the opportunity to write this book as part of a new book series on drying. Allison Shatkin from the publisher, Taylor & Francis, carefully handled this book project and provided very crucial editorial support.

There are many people, who I have not mentioned, who indirectly contributed to my CFD journey leading to this book. Special thanks to Professor Wu Zhonghua (Tianjin University of Science and Technology) who got me started on UDF coding in FLUENT. I thank Professor Timothy Langrish and Professor David Fletcher for hosting my brief visit to the University of Sydney in 2009 and for introducing me to the ANSYS Workbench package.

I thank my family for their continuous support of my endeavors. Lastly and most importantly, I thank my wife Michelle. Her encouragement and work alongside me for long weekends in the study provided me with the energy to complete this book.

Author

Meng Wai Woo is currently a senior lecturer of chemical engineering in Monash University, Australia. His research interest is in spray drying. He has experience in computational fluid dynamics (CFD) analysis of the spray-drying process and in applying this technique for industry. Within the area of spray drying, he is also examining the droplet evaporation, particle formation or interaction phenomena and in engineering the functionality of particles. Dr. Meng Wai Woo also explores new approaches to spray drying in introducing the antisolvent vapor precipitation approach and most recently, the narrow tube spray-drying technique.

Chapter 1

Introduction

The computational fluid dynamics (CFD) technique is already
well established and many books can be found describing
this technique in great detail. Implementing the technique for
modeling the spray-drying process, however, is not a trivial
endeavor. The engineer or modeler will have to understand
how to effectively numerically capture the important airflow
characteristics, droplet drying behavior, and the particles transport phenomenon typically occurring within the drying chamber. This is precisely the aim of this book: how to numerically
capture the important physical phenomena within a spray-
drying process using the CFD technique. Therefore, this book
is not a basic treatment of the CFD technique. It is also not
a discussion fundamentally focusing on the physics of spray
drying. The key word here is "numerical capture" and it is
the aim of this book to link the physical phenomena involved
with the numerical aspect on how they can be represented
numerically in a CFD simulation. The book will also include
numerical strategies in effectively capturing these phenomena
numerically. These strategies are collated from research work
in this area and from CFD industrial consultation particularly
to the dairy industry.

It will be noted later on that some parts of this book will make reference to the FLUENT software, which is part of the ANSYS workbench package. These elements were included mainly to highlight how some of the concepts discussed in this work are available in commercial CFD codes. To the industrial engineer, this may provide a reference on the accessibility of some of the numerical capabilities and the degree of work required to implement them. The author is by no means affiliated with ANSYS and this information is provided solely due to the familiarity of the author. There are numerous commercial CFD packages and open source codes available for the industry.

This book will be of interest to engineers and scientists interested in setting up their own CFD model of a spray-drying process. Apart from teaching the reader on how to set up their models, the discussion in this book will help the reader to identify the capabilities (or the uncertainty) of the CFD technique for spray drying. This will be particularly useful when interpreting the results from such simulations. Although basic knowledge of the spray-drying process is not necessary in reading this book, some exposure to or experience in the CFD technique will allow better appreciation and application of the strategies outlined in this book.

1.1 Why Do We Need CFD Simulation of Spray Drying?

Spray drying is widely used in the food, pharmaceutical, and detergent industry to produce dry powders. This is indeed a complex process and is highly nonlinear owing to the combination of droplet or particles transport within the drying chamber and the simultaneous moisture removal process. The control of these simultaneous phenomena, culminating in the different drying history experienced by the droplet or particle, is the key in ensuring reliable operation of a spray dryer. Therefore, in contrast to the traditional approach of treating the spray dryer as a black box, CFD simulation is now often used to provide a better understanding and visualization of the process. Such understanding can be very useful in designing new spray dryers or troubleshooting existing operations.

1.2 Components of CFD Simulation of Spray Drying

The core of a CFD simulation of a spray dryer, in fact for any CFD simulation, is the airflow or fluid flow prediction framework. This is numerically undertaken by solving the momentum equation of the air corresponding to the geometrical shape of the spray dryer and the operating conditions for the spray dryer, the boundary conditions of the simulations. Building on top of the fluid flow prediction, submodels are then incorporated into the larger CFD framework. These submodels are key numerical components, which distinguish the spray-drying simulation from the conventional fluid-only simulations. These submodels are shown in Figure 1.1.

All the submodels shown are important as they represent different physical aspects of spray drying. If possible, and for higher level of accuracy, it will be desirable to include all these submodels into the CFD simulation. It should be noted that each of these submodels introduces a certain level of uncertainty as they are normally developed independently and validated for specific applications, which may not be spray-drying applications. At the time of preparing this book, not all the submodels are fully well established or validated. The Lagrangian particle-tracking submodel is typically used to track and predict how particles move in the spray chamber. As part of the Lagrangian particle tracking, atomization submodels are then used to define the initial and boundary conditions of the particle tracking. The initial condition of the particle should correspond to the atomization process of the droplet and the atomization submodel accounts for this. If a particle impacts the wall during the simulation, the particle deposition submodel is then used to determine the boundary condition of the outcome of the particle–wall interaction. The key challenge is how to account for particle stickiness or adhesive characteristics in the deposition submodel. Once the movement of the droplet or particle is tracked in the simulation, the single droplet drying submodel will then be used to account for heat and mass transfer to and from the droplet. Development of this branch of submodel has attracted most of the attention in the development of spray-drying modeling as this forms the core of spray drying.

The above-mentioned submodels are commonly found and are "standard" requirement in all the spray-drying simulations reported. Some recent work has introduced the agglomeration submodel into the larger simulation framework. In comparison to the other submodels mentioned above, the agglomeration submodel is relatively more complex and it can be further broken down into several submodels combined to predict agglomeration of particles.

It will be noted that other quality development submodels are separated from the "core" submodels mentioned so far. This is because, in the current state of the models, the submodels are mainly used and implemented as a postprocessing step and do not feedback into the computation of the other submodels or the core fluid computation framework. Further development in these models may forge two-way couplings between these and the other submodels in the future, if required.

This book is organized based on individual components of a CFD simulation of spray drying as illustrated in Figure 1.1. Chapters 2 and 3 of this book focus on the core CFD airflow computation. The remaining chapters then focus on individual submodels. A theoretical framework is firstly given to the

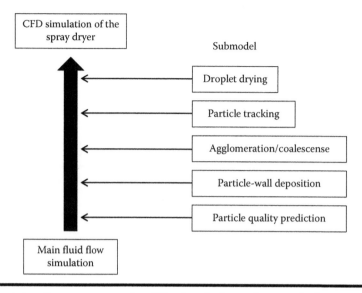

Figure 1.1   Components of a CFD simulation of spray drying.

reader followed by important considerations when implementing these submodels numerically.

These are the theoretical frameworks of a CFD simulation for spray drying. In most commercial CFD programs, the core airflow prediction and the Lagrangian particle tracking submodel are normally available as built-in features of the software. The other submodels may require additional coding and separate implementation into the commercial codes. For FLUENT, this refers to the use of the user-defined functions, whereas for CFX, this points to the use and implementation of the CCL codes.

Chapter 2

Basics of CFD

The aim of this section is to provide a brief overview of what a CFD simulation actually is. To put this into layman's terms, when we solve a CFD model or undertake a CFD simulation, what are we solving for? For many engineers who have seen or been shown a CFD simulation visualization, a question that may arise might be: how was the flow field generated? This section aims to provide a brief explanation to answer this question. Most books on CFD will touch on the equations used in the model. This book, however, focuses on the concept and the overall idea of CFD. This will be particularly useful for engineers who have not undertaken an advanced course in fluid flow or numerical modeling. Therefore, the theoretical aspects or the equations involved will be covered only briefly. The theoretical aspects of fluid mechanics or transport phenomena essential in CFD are covered in typical undergraduate or in advanced engineering courses. Even if one is familiar with the theoretical aspects, this chapter may give a "practical" aspect of the theory. The focus is mainly to familiarize the readers with the jargon and the basic concept behind CFD simulations, which will be essential for further reading in this book. For simplicity, only a discussion of a two-dimensional aspect of CFD simulations will be presented; however, the concept discussed can be easily extended to three-dimensional or to different coordinate reference frameworks. This section, however, is not a complete theoretical treatment of the CFD technique. More detailed literatures on the basic of numerical solutions can be found from the references provided (Patankar 1980; Versteeg and Malalsekera 2007). The basis of the "teaching approach" of this section draws inspiration mainly from the work of Patankar (1980).

2.1 How Are Those Velocity and Temperature Plots Generated?

At the core of a CFD simulation is the solution of the momentum equation of a fluid, which may be any fluid, within a specific simulation domain. For fluid that is Newtonian and is incompressible, conditions that are typically encountered in spray drying, the Navier–Stokes equation (which is a modification from the basic momentum equation) is typically used. For a two-dimensional system in the Cartesian framework it has the form,

$$\frac{\partial \rho v_x}{\partial t} + v_x \frac{\partial \rho v_x}{\partial x} + v_y \frac{\partial \rho v_x}{\partial y} = \mu \frac{\partial^2 v_x}{\partial x^2} + \mu \frac{\partial^2 v_x}{\partial y^2} - \frac{\partial P}{\partial x} + \rho g_x \qquad (2.1)$$

$$\frac{\partial \rho v_y}{\partial t} + v_x \frac{\partial \rho v_y}{\partial x} + v_y \frac{\partial \rho v_y}{\partial y} = \mu \frac{\partial^2 v_y}{\partial x^2} + \mu \frac{\partial^2 v_y}{\partial y^2} - \frac{\partial P}{\partial y} + \rho g_y \qquad (2.2)$$

It can be seen that in the momentum equation, each term carries the unit of the rate of the change of momentum per unit volume of the fluid. The differential form of the equation denotes that this is a continuous equation, which can be used to describe the velocity of the fluid at any specific point within a simulation domain. Imagine water flowing into a pipe in a two-dimensional manner and leaves the pipe at the outlet (Figure 2.1) and consider a single random point within the system. The time

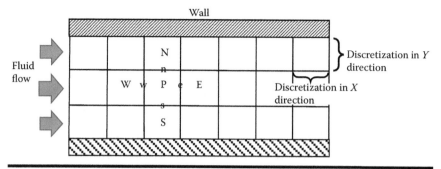

Figure 2.1   Discretization of the simulation domain and the notation used.

derivative on the left-hand side of the equations denotes rate of change of momentum at that particular point. The second and third term on the left-hand side of the equations denote the change in momentum across that point in the X and Y direction due to bulk convective flow of the fluid in the system. As the fluid flows, however, momentum may tend to diffuse or spread owing to shearing between regions of fluid with different velocities. Imagine a jet of high-velocity air coming out from a nozzle, which quickly expands and diffuses. This phenomenon in momentum transfer is denoted by the double derivative terms on the right-hand side of the equations. The pressure derivative denotes the differential pressure "driving" the fluid and the gravity term denotes the effect of gravity on fluid flow.

The equations imply that momentum change in the simulation domain can be mainly denoted by the velocity change in the system. Therefore, when we solve the momentum equation, from an application point of view, we are essentially solving or finding out how velocity changes at each point in the simulation domain. In other words, we are finding out the velocity at each point in the simulation domain. This view point holds even if the density varies as momentum changes are mainly determined by the change in velocity. How do we then solve this equation?

For very simple two-dimensional systems, the momentum equations can be simplified and with calculus manipulation, analytical solutions for the change in velocity can be obtained.

The term "simple" is strictly relative and the spray-drying system will be "too complex" for an analytical solution. Therefore, the equation will have to be solved numerically. The first step is to discretize or "chop up" the simulation domain into tiny control volumes. As shown in Figure 2.1, the rectangular system was arbitrarily discretized into 21 control volumes.

The notation used here follows an adaptation from Patankar (1980). The next step is to discretize the momentum equation into algebraic equations. Let us consider discretizing the momentum equation at one particular finite volume P in order to find the velocity of the fluid at that particular element volume, as shown in Figure 2.2. To describe the velocity at elemental volumes adjacent to that particular elemental volume, the subscript notations as shown in Figure 2.2 are used.

For simplicity, let us firstly assume a steady-state fluid flow that eliminates the time derivative term on the left-hand side of the equation. Let us consider the differential equation for the x-component of the velocity field,

$$v_x \frac{\partial \rho v_x}{\partial x} + v_y \frac{\partial \rho v_x}{\partial y} = \mu \frac{\partial (\partial v_x / \partial x)}{\partial x} + \mu \frac{\partial (\partial v_x / \partial y)}{\partial y} - \frac{\partial P}{\partial x} + \rho g_x$$

(2.3)

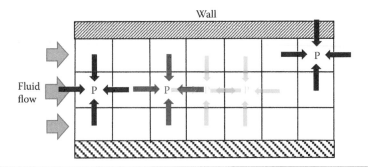

Figure 2.2 Implementation of the discretized Navier–Stokes equation.

With the control volume approach (Versteeg and Malalsekera 2007), typically used by many commercial codes nowadays, the next step to discretizing the equation is by integrating the differential equation over a particular control volume,

$$\int_{CV} \rho v_x \frac{\partial v_x}{\partial x} dCV + \int_{CV} \rho v_y \frac{\partial v_x}{\partial y} dCV$$

$$= \int_{CV} \mu \frac{\partial(\partial v_x/\partial x)}{\partial x} dCV + \int_{CV} \mu \frac{\partial(\partial v_x/\partial y)}{\partial y} dCV - \int_{CV} \frac{\partial P}{\partial x} dCV$$

$$+ \int_{CV} \rho g_x dCV \tag{2.4}$$

The integrals on the left-hand side of the equation are volume integral of the divergence of the velocity and density product. Similarly, the first two terms on the right-hand side of the equation represent the volume integral of the divergence of the *x*-velocity gradient. From the divergence theorem, the volume integral of the divergence of a vector can be represented by the surface integral of the vectors, further modifying the integral of the equation,

$$\int_A \rho v_x (v_x \cdot \vec{n}) dA + \int_A \rho v_y (v_x \cdot \vec{n}) dA$$

$$= \int_A \mu \left(\frac{\partial v_x}{\partial x} \cdot \vec{n} \right) dA + \int_A \mu \left(\frac{\partial v_x}{\partial y} \cdot \vec{n} \right) dA$$

$$- \int_{CV} \frac{\partial P}{\partial x} dCV + \int_{CV} \rho g_x dCV \tag{2.5}$$

The $\vec{n}$ term is the normal vector corresponding to the surface of the area integration. Considering a discrete control volume, small changes in the *x*, *y*, and *z* spatial lengths can then be substituted into the equation to arrive at the discrete form of the continuum form of the equation,

$$
\rho v_{x,n}(v_{x,n}\cdot\vec{n})\Delta x + \rho v_{x,s}(v_{x,s}\cdot\vec{n})\Delta x + \rho v_{x,e}(v_{x,e}\cdot\vec{n})\Delta y
$$
$$
+ \rho v_{x,w}(v_{x,w}\cdot\vec{n})\Delta y + \rho v_{y,n}(v_{x,n}\cdot\vec{n})\Delta x + \rho v_{y,s}(v_{x,s}\cdot\vec{n})\Delta x
$$
$$
+ \rho v_{y,e}(v_{x,e}\cdot\vec{n})\Delta y + \rho v_{y,w}(v_{x,w}\cdot\vec{n})\Delta y
$$
$$
= \mu\left(\left.\frac{\partial v_x}{\partial x}\right|_n\cdot\vec{n}\right)\Delta x + \mu\left(\left.\frac{\partial v_x}{\partial x}\right|_s\cdot\vec{n}\right)\Delta x + \mu\left(\left.\frac{\partial v_x}{\partial x}\right|_e\cdot\vec{n}\right)\Delta y
$$
$$
+ \mu\left(\left.\frac{\partial v_x}{\partial x}\right|_w\cdot\vec{n}\right)\Delta y + \mu\left(\left.\frac{\partial v_x}{\partial y}\right|_n\cdot\vec{n}\right)\Delta x + \mu\left(\left.\frac{\partial v_x}{\partial y}\right|_s\cdot\vec{n}\right)\Delta x
$$
$$
+ \mu\left(\left.\frac{\partial v_x}{\partial y}\right|_e\cdot\vec{n}\right)\Delta y + \mu\left(\left.\frac{\partial v_x}{\partial y}\right|_w\cdot\vec{n}\right)\Delta y - \frac{\partial P}{\partial x}\Delta x\Delta y + \rho g_x\Delta x\Delta y \quad (2.6)
$$

It should be noted that for the two-dimensional system considered, the length scale in the z-direction is of unit length 1. Therefore, the Δz term was omitted from all the terms in the equation above. Conveniently, developing this in the Cartesian framework, terms in the above equation with the velocity flow field $\rho v_{x,i}$ in parallel with the discrete surface can then be equated with zero and be omitted as the divergence theorem sums only the integral of the flow field normal to the system boundary surface. It should be noted that the location of velocities at the surface is denoted by lower case letters corresponding to the north, south, east, and west direction relative to the P control volume. This is shown in Figure 2.1.

The simplified equation then takes the following form,

$$
\rho v_{x,e}(v_{x,e})\Delta y + \rho v_{x,w}(v_{x,w})\Delta y + \rho v_{y,n}(v_{x,n})\Delta x + \rho v_{y,s}(v_{x,s})\Delta x
$$
$$
= \mu\left(\left.\frac{\partial v_x}{\partial x}\right|_e\right)\Delta y + \mu\left(\left.\frac{\partial v_x}{\partial x}\right|_w\right)\Delta y + \mu\left(\left.\frac{\partial v_x}{\partial y}\right|_n\right)\Delta x
$$
$$
+ \mu\left(\left.\frac{\partial v_x}{\partial y}\right|_s\right)\Delta x - \frac{\partial P}{\partial x}\Delta x\Delta y + \rho g_x\Delta x\Delta y \quad (2.7)
$$

The $\rho v_{x,e}$, $\rho v_{x,w}$, $\rho v_{y,n}$, and $\rho v_{y,s}$ terms represent the mass flow flux at the face of the control volume delineating the convection of fluid "transporting" the momentum in the flow field. This presents nonlinearity into the solution. The solution of the flow field will thus require an initial guess of the flow field followed by successive iterations to achieve a correct flow field. Following Patankar (1980), let us denote these terms as $F_{x,n}$, $F_{x,w}$, $F_{y,n}$, and $F_{y,s}$, respectively,

$$F_{x,e}(v_{x,e})\Delta y + F_{x,w}(v_{x,w})\Delta y + F_{y,n}(v_{x,n})\Delta x + F_{y,s}(v_{x,s})\Delta x$$

$$= \mu\left(\frac{\partial v_x}{\partial x}\bigg|_e\right)\Delta y + \mu\left(\frac{\partial v_x}{\partial x}\bigg|_w\right)\Delta y + \mu\left(\frac{\partial v_x}{\partial y}\bigg|_n\right)\Delta x + \mu\left(\frac{\partial v_x}{\partial y}\bigg|_s\right)\Delta x$$

$$-\frac{\partial P}{\partial x}\Delta x\Delta y + \rho g_x\Delta x\Delta y \tag{2.8}$$

It can be seen that the semi-discretized equation is now a function of the velocity and velocity gradient at the surfaces of the control volume. In the control volume approach, the flow field information is stored only within the control volume. This implies that surface properties have to be estimated from the information in the control volume.

There are many different forms of approximation ranging from a method involving first-order, second-order or an even higher order of accuracy. More details on this can be found in the literature (Patankar et al. 1980; Versteeg and Malalsekera 2007). For this example, we arbitrarily select the following for simplicity and for illustration purposes:

1. For the convective terms, the face velocity for the east and west terms follows the properties of the control volume upstream. The face velocity for the north and south terms takes the average of the velocity values of the control volumes adjacent to the face.
2. The velocity gradient was estimated from the difference in the velocity at the adjacent control volumes.

The equation then takes the following form,

$$F_{x,e}(v_{x,E})\Delta y + F_{x,w}(v_{x,P})\Delta y + F_{y,n}\left(\frac{v_{x,N}+v_{x,P}}{2}\right)\Delta x$$

$$+ F_{y,s}\left(\frac{v_{x,S}+v_{x,P}}{2}\right)\Delta x$$

$$= \mu\left(\frac{v_{x,E}-v_{x,P}}{\Delta x}\right)\Delta y + \mu\left(\frac{v_{x,P}-v_{x,W}}{\Delta x}\right)\Delta y + \mu\left(\frac{v_{x,N}+v_{x,P}}{\Delta y}\right)\Delta x$$

$$+ \mu\left(\frac{v_{x,P}+v_{x,S}}{\Delta y}\right)\Delta x - \frac{P_E - P_W}{2}\Delta x\Delta y + \rho g_x \Delta x \Delta y \qquad (2.9)$$

Rearranging this equation,

$$v_{x,P} = \frac{\begin{bmatrix}((\Delta y\mu)/\Delta x)v_{x,E} + F_{x,e}(v_{x,E})\Delta y\end{bmatrix} \\ +\begin{bmatrix}((\Delta x\mu)/\Delta y)v_{x,N} + ((F_{y,n}\Delta x)/2)v_{x,N}\end{bmatrix} \\ +\begin{bmatrix}((\Delta x\mu)/\Delta y)v_{x,S} + ((F_{y,s}\Delta x)/2)v_{x,S}\end{bmatrix} \\ -\begin{bmatrix}((\Delta y\mu)/\Delta x)v_{x,W}\end{bmatrix}-\begin{bmatrix}((P_E - P_W)/2)\Delta x\Delta y + \rho g_x \Delta x\Delta y\end{bmatrix}}{\begin{bmatrix}F_{x,w}\Delta y + ((F_{y,n}\Delta x)/2) + ((F_{y,s}\Delta x)/2)\end{bmatrix}}$$

$$(2.10)$$

From this example, it can be seen that the velocity of fluid flow at the central elemental volume is actually described by the velocity and pressure of the adjacent elemental volumes. Therefore, if we "drag" this equation across to different elemental volumes across the simulation domain, we can then describe the velocity of the fluid flow at each discretized volume or space in the simulation domain (red and green sets in Figure 2.2). However, at the elemental volume next to the wall or adjacent to the inlet or outlet point, we may encounter a discrepancy in which the velocity at the wall or velocity at the inlet or outlet is then required (purple sets in Figure 2.2). The additional details, which are in essence not part of the internal of the simulation domain, are called boundary conditions and are input parameters into the simulation. The effect of different values then pervades into the simulation, affecting the prediction of the velocity. In the example above, a very simple discretization scheme was employed and a steady-state solution was shown. For transient simulations, a similar discretization approach with an addition of the time integral can be added.

The prediction of the flow field, regardless of whether it is a transient or steady-state simulation, is also affected by the $F_{x,n}$, $F_{x,w}$, $F_{y,n}$, $F_{y,s}$ terms and the pressure at each control volume. These are the initial values to the simulation. If the pressure field is specified on the same control volume in which the velocity is solved, there is potential for the prediction of an unrealistic flow field should a wavy pressure field be specified. This is known as the "checkerboard" effect and the mathematical description of this can be found in Patankar (1980). For this reason, the velocity field and the pressure field (also that of the continuity equation) are solved on separate staggered grids. Detailed mathematical treatment of this is given in the literature (Patankar 1980).

If successive iterations are undertaken, the velocity field generated from the model can then be used as the new guess of the velocity flow field for the subsequent iteration, with the possibility that successive iterations will eventually lead to a good approximation of the physical flow field. However, such numerical pressure correction is not implied in the numerical formulation in Equation 2.10. Therefore, a pressure–velocity coupling scheme is required to iteratively correct the guess of the pressure field, which subsequently also corrects the velocity field as differential pressure physically "drives" the velocity. Details on such pressure–velocity coupling are not covered here and more information can be found in the literature (Patankar 1980).

It is hoped that the example above will give an idea on how the velocity vector plots are generated in a CFD simulation. The properties used are mainly affected by the temperature of the fluid. Determining suitable temperature will require the similar solution of the energy equation given below for a two-dimensional system.

$$\frac{\partial \rho h}{\partial t} + \frac{\partial \rho v_x h}{\partial x} + \frac{\partial \rho v_y h}{\partial y} = k\frac{\partial^2 T}{\partial x^2} + k\frac{\partial^2 T}{\partial y^2} + S_h \qquad (2.11)$$

The equation above may need to be modified for energy transfer due to water vapor transport (in the context of spray dryers) and energy generation due to turbulent viscous dissipation or due to the drying process.

2.2 How Is Turbulence Captured?

A complete theoretical treatment on turbulent flows is beyond the scope of this book. Engineers entering the field with a basic degree may only be exposed to a brief study of turbulence. Detailed mathematical treatment of turbulence may only be covered at a more advanced level of study. This part of the chapter is aimed at providing a simple phenomenological overview of turbulence modeling targeted at readers with minimal knowledge of turbulence modeling and the mathematics involved. Figure 2.3 illustrates a stream of fine mist jetting out from a pipe with transient turbulent structures.

It can be seen that in general, at different times the smoke is moving in an outwards direction from the pipe. However, "superimposed" in the average outwards direction are visible transient turbulent eddies which are generated by shearing between the jet and the ambient adjacent air. These eddies may vary in size. Depending on the turbulence of the overall flow, the time scale of these eddies may also vary. Therefore, numerical capture of fluid flow turbulence is complex because of the different length scales and time scale of these eddies.

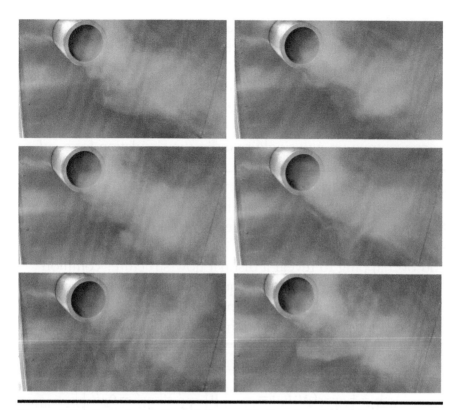

Figure 2.3 Snapshots of an airflow with fine droplets delineating the multiscale transient turbulent structures. (From Mansouri, S. et al. Narrow tube spray drying. *Drying Technology* **2016, 34(9), 1043–1051.)**

From the preceding section, by utilizing a spatial (and possibly time) discretization approach to the numerical solution, a limitation is then imposed on the smallest length scale and time scale of turbulence which can be explicitly captured in the simulation for turbulence modeling. One way to overcome this is to discretize the simulation domain and the time domain into excessively small elements so that the smallest and most fundamental level of turbulence is captured. These are called direct numerical simulations (DNS). However, due to the excessive computation requirements, DNS may have limited application for routine engineering use.

In contrast, another approach is to model any turbulence smaller than the length scale of the discretized simulation domain with a suitable mathematical description. In order words, there is no need to discretize the simulation domain to excessively small length scales. A common approach widely used is the Reynolds average Navier–Stokes (RANS) approach. The basis for the RANS approach is ensemble averaging. A simple way to imagine ensemble averaging is that if you were to concentrate on any single point in the simulation for a sufficiently long time, although the flow may be fluctuating in direction and magnitude due to turbulence, there will be an average representative velocity magnitude and direction of the flow at that point—the ensemble averaged flow field. One can imagine making all six pictures in Figure 2.3 semi-transparent and overlapping to reveal a distinct flow structure "common" at each flow time. The RANS equation, which in essence is the ensemble average of the momentum equation, is given below and the solution of this equation then provides the ensemble averaged flow field,

$$
\frac{\partial \rho v_i}{\partial t} + \frac{\partial}{\partial x_j}(\rho v_i v_j) = \frac{\partial}{\partial x_j}\left(\mu\left(\frac{\partial v_i}{\partial x_j} + \frac{\partial v_j}{\partial x_i} - \delta_{ij}\bar{P}\right)\right)
$$
$$
+ \frac{\partial}{\partial x_j}(-\rho v_i v_i) - \frac{\partial P}{\partial x_i} \tag{2.12}
$$

It can be noticed that in the ensemble averaging process, a new term, which are the Reynolds stresses, needs to be solved (the second term on the right-hand side of the equation). Solution of the Reynolds stresses is very often referred to as providing closure to the RANS equation. Most of the different models available in CFD simulations pertain to the different approaches to solve the Reynolds stresses. This will be covered in greater details in Section 3.1. In general, the large-scale eddy (LES) simulation approach in essence also provides a solution of the ensemble averaged flow field. However, this approach resolves the turbulence to a finer length scale and makes different assumptions on the fluctuating component of the fluid.

2.3 Common Basic Numerical Strategies

Throughout this book, numerical strategies will be included specific to different parts of the modeling of spray dryers. A few strategies for general basic fluid flow simulation are included here for readers with minimal experience with the CFD technique. From the discussion in Section 2.1, it can be seen that the discretized solution of the flow field is partly affected by the size of the discretization of the simulation domain. Therefore, it is essential that mesh independence testing be undertaken for any CFD simulation. In essence, the finer the discretization, the more accurate is the simulation. However, too fine a mesh will result in the need for excessive computational resources. On the other hand, too coarse a mesh size will lead to nonphysical flow prediction. One should start the simulation with a grid size sufficient to resolve key important flow structures in the simulation and then progressively refine (making the mesh size smaller) at regions of high velocity or temperature gradient. This can be undertaken repeatedly from successive simulations until a flow field solution is independent of further refinement to the system. Another commonly adopted practice is to use only second-order (or higher) linearization of the partial different equations in a CFD simulation. First-order linearization should only be used to provide an initial "look" of the simulated flow field.

Chapter 3

Airflow Modeling

3.1 Turbulence Model Selection

A survey in the literature on CFD modeling of spray dryers will reveal that early initial work in this area utilizes the RANS with closure equations to model the turbulence of the flow (Oakley and Bahu 1993; Harvie et al. 2002) or solving the individual components of the RANS stress tensors (Oakley and Bahu 1993; Huang et al. 2004). The turbulence parameter (e.g., turbulent kinetic energy) is then used in the prediction of the particle dispersion phenomenon. Usage of the closure equations approach provides a convenient approach, finding a compromise between the computational requirements and the level of details in the turbulence prediction. It should be noted that a compromise in the "level of detail" may not necessarily mean compromising the accuracy of the turbulence prediction.

There are relatively few experimental measurements of the turbulence within the spray chamber; internal measurements in spray dryers are relatively scarce. Therefore, validation of the flow fields predicted and reported in the literature so far is mainly done by comparing with the measured velocity flow field within a spray chamber. The difficulty in making such measurements is discussed in Chapter 7.

Why is accurately capturing the modeling of the turbulence important? The most direct effect of the turbulence model, which is obvious, is on the prediction of the velocity flow field. Secondly, which may be less obvious for CFD users, many of which may have dealt primarily with fluid-only simulations, is that the turbulence parameters in the flow field also affect the prediction of particle dispersion, particularly for the smaller particles and also near-wall dispersion, which subsequently affects particle-wall deposition prediction; the latter is highly dependent on the model used. This will be discussed in Chapters 4 and 6, respectively. Thirdly, the degree of turbulence, particularly at the atomizer region, will also affect and can be used to intensify the drying process (Southwell et al. 1999).

Validating with observed experimental observations in a series of work, the RANS approach (specifically the $k - \varepsilon$) was found to be able to capture the transient behavior of the flow field in the spray chamber. The transient behavior in discussion refers to the self-sustained oscillating flow behavior in the chamber, which will be discussed in great detail in the next section (Guo et al. 2003). This capability of the $k - \varepsilon$ approach was evaluated for both conditions with and without swirls in the air. In addition, for the short-form spray dryers, the RNG version of the $k - \varepsilon$ model was suggested to provide the closest match to the measured experimental flow field from the chamber (Huang et al. 2004).

There is a trend observed in the field of spray-drying CFD simulations in utilizing turbulence models with a higher level of details. One such report uses a scale-adaptive simulation approach, which provided more details on the vortices occurring within the spray chamber (Fletcher and Langrish 2009). In addition to that, there were also several reports utilizing the LES approach to model the turbulence in an industrial-scale spray dryer (Jongsma et al. 2013). From the LES simulations, an interesting numerical prediction was that the coherent jet fluctuation structures predicted with the RANS approach was not observed; fluctuations became more chaotic.

Although the use of a more detailed simulation of the turbulence will certainly provide more information on the flow structures, at the moment it remains unclear whether such a high level of detail significantly affects or to what extents it affects the particle drying history or particle trajectory prediction within the spray chamber. This is because, to the best of the knowledge of the author, the higher detailed simulations described above were mainly undertaken without the incorporation of particles or droplets. A systematic comparison between the RANS with the closure equations approach to the high-level detail turbulence modeling approach is certainly required.

3.2 Transient Flow versus Steady Flow

3.2.1 In What Situations Do We Expect Transient Flows?

It is important to clearly define what is meant by transient flow in the context of spray drying. The term "transient" is used to define coherent (but may be chaotic) large-scale fluctuations in the airflow pattern within the chamber (Guo et al. 2003; Southwell and Langrish 2000; Fletcher et al. 2006). The fluctuation behavior is more clearly visualized by the flapping or self-sustained oscillation of the central jet of airflow in the radial direction of the chamber. The self-sustained oscillation of the central jet also results in oscillating flow reversals between upward flowing air and downward flowing air in the regions away from the central airflow region (typically downwards for a cocurrent spray dryer). Figure 3.1 illustrates such self-sustained fluctuations as a series of subsequent snap shots of the flow field from a spray dryer simulation. A very important feature of this fluctuating behavior is the three-dimensional nature of the fluctuation. This three-dimensionality can be clearly interpreted by examining the flow field from the horizontal cross section of the drying chamber (Figure 3.1).

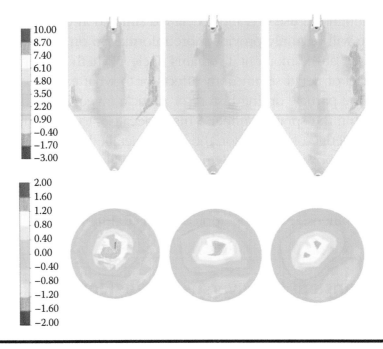

Figure 3.1   Example of the self-sustained oscillating flow in a spray dryer. The circular contour plot is the top view of the flow field just below the cylinder–cone interface. The contour represents the axial velocity in m/s (13 s interval between the first and second snapshots and 10 s between the second and third snapshots in the horizontal direction).

Earlier CFD work in spray drying, before consideration of this phenomenon, utilizes steady-state simulations in the CFD framework (Oakley and Bahu 1993; Kieviet and Kerkhof 1997). With this assumption, these earlier reports were mainly steady-state two-dimensional simulations. Furthermore, the steady-state two-dimensional simulations also resulted in a converged solution, which may suggest predominantly physical steady-state flows, otherwise solving a transient problem in a steady-state solution may result in nonconvergence numerically. One possible reason, as suggested by Langrish and coworkers (Fletcher et al. 2006), is that the two-dimensional simulation domain might actually "restrict" the possible transient behavior in the airflow, resulting in the steady-state convergent results.

Such a numerical restriction may also be found in axisymmetric simulations or partially three-dimensional simulations with a "wedge" of the simulation domain using periodic boundary conditions. Therefore, it will be important to use a full three-dimensional simulation to allow any possible fluctuating flow field to be effectively captured. Nevertheless, it is important to note that although the transient behavior was not captured by these simulations, they did provide a reasonable match to the available experimental data presented at that time. It is unclear at the moment to what extent the transient airflow behavior affects the drying behavior and most importantly the residence time of the particles in the drying chamber.

What is the reason for the coherent long-scale transient flow field behavior? A series of investigations focused on the instability when a jet of air undergoes sudden pipe expansion (Guo et al. 2001a,b). Hot airflow into a spray dryer, particularly if a central air inlet is used, is analogous to such an expansion; that is, air from a relatively smaller inlet going into a relatively larger chamber. Under such sudden expansions, part of the central jet tends to "split" and recirculate backwards towards the inlet region owing to the low-pressure regions in the pipe generated by the central jet. As a result of such recirculations, the central jet will then flap in a coherent or chaotic manner, depending on the expansion ratio and the inlet air velocity.

When sudden constrictions are introduced to this study, it was found that the constriction typically affects the stability of the airflow as airflow blocked by the constriction will tend to recirculate within the chamber across to opposing sides of the central jet, causing self-sustained oscillation. Figure 3.2 illustrates this phenomenon, which for the remainder of this chapter will be called the "jet feedback" mechanism. It should be noted that the jet feedback mechanism causes the jet to flap in a three-dimensional manner about the central axis of the spray chamber. Such flapping was observed even in larger scale spray dryers experimentally and numerically, even

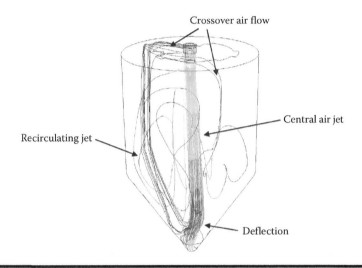

Crossover air flow

Central air jet

Recirculating jet

Deflection

Figure 3.2   Jet feedback mechanism.

for dryers that have air outlets at the top (air is recirculated upwards with a bottom fluidized bed) or with the air outlet at the bottom (Gabites et al. 2010; Jongsma et al. 2013).

Swirling of the inlet air is used in some spray dryers to enhance the heat and mass transfer between the droplets and the air. Numerical and experimental simulations suggest that the incorporation of swirls tends to produce a more predictable transient flow behavior (Southwell and Langrish 2000, 2001). In simulations on a pilot-scale dryer, another distinct feature of the inlet induced swirls is the expansion of the central jet owing to the rotation of the flow. This will then lead to the breakdown and splitting of the central vortex jet depending on the angle of the swirl. Similar observations in the vortex breakdown can be observed even for spray dryers with atomizer-induced swirls (Woo et al. 2012). There are more important details such as the flow fluctuation frequency and the direction and motion of the flapping behavior of the central jet, which will not be covered in this chapter; this information can be accessed via the references cited and will be useful for more detailed comparison with any industrial work. For a general illustration, Figure 3.3 shows some of the

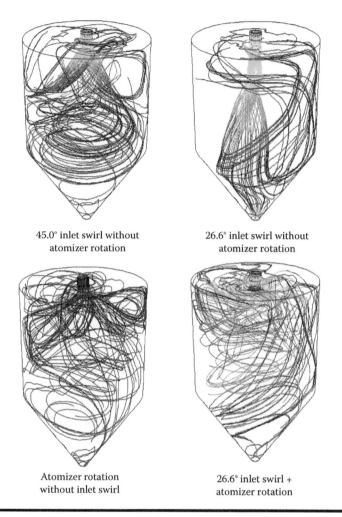

45.0° inlet swirl without
atomizer rotation

26.6° inlet swirl without
atomizer rotation

Atomizer rotation
without inlet swirl

26.6° inlet swirl +
atomizer rotation

Figure 3.3   Illustrations on the effect of swirls on the flow field pattern in a spray dryer.

typical flow features that may be expected by the incorporation of inlet swirls and rotating atomizer-induced swirls into a spray dryer simulation.

Digressing from the discussion on the transient behavior of flow, there is also a school of thoughts that swirls increase the residence time of the particles, providing a longer time for drying before the particles leave the system. A good example of this strategy in place is in spray towers for nonheat-sensitive materials such as detergents. However, recent evidence from an industrial-scale tower showed that the swirling flows actually enhance deposition of the particles on the wall. Subsequent reentrainment of these particles emerged as indistinguishable products that left directly without depositing (Francia et al. 2015). In the opinion of the author, it is unclear to what extent this contributes to the "extra residence time" for drying.

In general, from the author's experience, most spray dryers predominantly exhibit transient oscillating airflow behavior. However, in certain conditions, the flow field may exhibit steady-state behavior (Woo et al. 2009a). As described earlier, the possible mechanisms that generate the transient behavior are the (1) instability of the jet when undergoing a sudden expansion from the air inlet to the chamber geometry and (2) the feedback of the jet owing to the outlet constriction which causes an oscillating pressure imbalance which pushes the central jet. Two independent reports have shown that if the outlet constriction is significantly further from the inlet of the jet (long tower-like spray chamber), the transient behavior is not observed numerically (Guo et al. 2003). In addition, a wider chamber diameter relative to the size and velocity of the jet also resulted in steady-state airflow behavior (Woo et al. 2008a,b,c,d). From the latter report, it was deduced that if sufficient space in the chamber is provided to "equalize" the pressure in the region surrounding the central jet, the feedback effect mentioned earlier might be reduced, resulting in the diminishing of the flapping. This effect, however, will need to be evaluated relative to the inlet central jet air velocity as higher velocity will induce more intense feedback from the constriction and vice versa. Being aware of such possible steady-state conditions is very important, as it will significantly affect the required simulation time. Strategies to handle both steady state and transient flow behavior will be discussed in the next section.

3.2.2 Important Numerical Strategies for Transient Flows

There are a few strategies introduced here, which can be adopted to model the self-sustained transient airflow behavior. The first numerical strategy is to start the simulation with a steady-state simulation. For the steady-state simulation, it is quite common to initialize the flow field adopting the velocity at the inlet of the spray dryer. It should be noted that such an initialization approach mainly patches the whole flow field with a "blanket value." The key is to achieve an initial flow field as close as possible to the actual flow field. While such an initialization is suitable for a tower-like spray dryer, it may not be suitable for a short-form spray dryer or for spray dryers in which the top inlet is recirculated upwards to leave the chamber. The author's preference, partly adopted as a consistent practice, is to initialize the flow field with a still-air condition. To be more precise, the humidity level and the temperature of the air are initialized to follow that of the expected ambient air conditions, while the velocity and turbulence components are initialized as zero akin to no airflow within the chamber. With respect to FLUENT, the hybrid initialization will also be a suitable approach for this. However, the stability of using the hybrid initialization has not been fully tested by the author at the time of writing.

The initial steady-state solution will result to two possible outcomes. If the flow field is inherently transient with self-sustained flapping behavior, the steady-state solution will only partially converge. A partially converged solution can be deduced by the presence of a nonsymmetric flow field. In fact, at subsequent iterations, the flow field will tend to change significantly (see Figure 3.4). This is the numerical artifact of trying to solve a predominantly transient flow field in a steady numerical framework. In addition, this is also reflected in significant and continuous oscillation of the flow field local

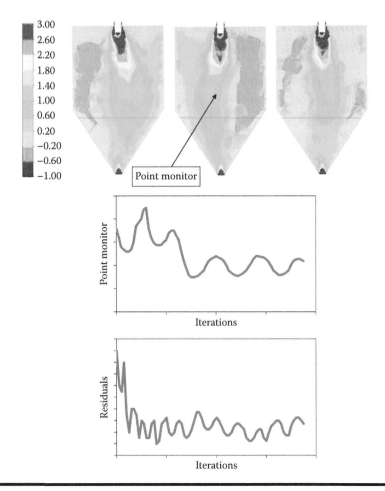

Figure 3.4 **Signs of a self-sustained transient flow field in the initial steady-state solution (the contour plot was generated for inlet air at high Reynolds number with 26.6° swirl—the three snapshots were obtained from different iterations).**

residue. An illustration of such an oscillating residue pattern is shown in Figure 3.4. It should be noted that, depending on the initial value of the flow field and the degree of transient behavior in the flow field, the residual may initially reduce significantly, giving a false indication of a normal steady-state solution, prior to reaching an oscillating pattern. For this reason, it will be important not to rely on the automatic check for convergence based on the residual, typically in commercial

CFD codes, in case the residuals dropped below the threshold before arriving at an oscillating pattern.

In contrast, if the flow field is predominantly steady state, the simulation will converge to a stable velocity flow field that is symmetrical (Figure 3.5). Corresponding to a steady-state solution, the residual of the simulation will also progressively

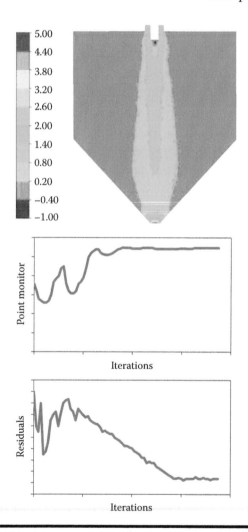

Figure 3.5 Signs of a predominantly steady-state conditions simulated for a spray dryer (the contour plot was generated for inlet air at low Reynolds number without swirl—the three snapshots were obtained from different iterations).

reduce and, given sufficient time, will reach a horizontal plateau pattern. However, if this happens, it will be very important to further verify the steady-state behavior by progressively making further refinements to the mesh. Although mesh refinement is typically undertaken to arrive at mesh-independent simulation results, the mesh refinement for steady-state verification has a different purpose. Fine mesh refinement is essential in capturing the potential transient behavior of the flow field. If the simulation consistently showed steady-state behavior even at higher mesh refinements, it can then be deemed to exhibit steady-state airflow behavior.

For the transient flow field delineated by the preliminary steady-state simulation, the next step is then to switch over to the transient simulation framework. The flow field from the initial steady-state simulation is then used as the initial solution to the transient simulations. In the transient framework, the flow field should then be allowed to numerically develop to reach a self-sustained transient behavior. Full development of the flow field can be monitored by continuous recording of the flow velocity at arbitrary locations within the chamber. Sustained oscillation of the velocity is an indication of the full development of the flow. It is important not to monitor too close to the inlet region of the airflow as this may give a false reading of a relatively fast flow development. The center axis of the chamber will be a suitable place to monitor the development of the oscillating air jet. Analysis and data collection from the simulation should only be undertaken once the flow field has achieved self-sustained oscillation; that is, once it is fully developed.

In using such steady-transient simulation strategy, the initial steady-state simulations will then serve to reduce the simulation time required to develop the transient flow as the steady state already provides a semi-developed flow field. For certain spray-drying configurations and operating conditions, sometimes such a strategy may not work because very aberrant flow field patterns may be observed from the initial steady-state solution. Using the aberrant flow field as the initial condition for the transient simulation may result in nonphysical results from the transient framework or even result in divergence of the numerical solution. It is unclear at the moment on how to "identify" such a case beforehand. The author speculates that this may happen for spray-drying chambers, which tend to have very intense or rapid coherent transient flow fluctuations.

For such cases, it will be important to start the simulation with the transient simulation framework right from the beginning. Such tough cases may require "step-by-step" buildup of the complexity in the simulation. The author's rule of thumb is to simulate the dryer based on how the dryer would have been started up in practice. Let us take a rotary atomizer-fitted, short-form spray dryer for example. The simulation domain can firstly be initialized with zero air velocity but initialized with the humidity and temperature of that of ambient air. The transient solution is then developed, incorporating the energy balance. Once the transient flow field is developed, rotation to the atomizer wall is then progressively increased and the transient flow field is allowed to numerically develop into a self-sustained transient behavior.

An additional note on the simulation time step size. Ideally, it will be essential to progressively reduce the time step size to find time step size independence. However, unlike the steady-state case in which the velocity profile at the same position within the simulation domain can be compared, it is relatively difficult to compare for a transient simulation. This is because the transient behavior is not a developing-type, but rather a self-sustained transient behavior; each simulation will require different time duration to achieve full flow development. Therefore, the velocity profile at any point in the simulation may not be the same even at the same simulation flow time. A rule of thumb commonly adopted is to ensure that the time step size is approximately two magnitudes smaller than the residence time of the airflow in the simulation domain.

The airflow residence time can be estimated by roughly dividing the main jet velocity within the chamber with the approximate length scale in which it will travel until it leaves the chamber. For such rough estimation, sometimes the highest approximated airflow velocity in the system can be used so that important features of the flow can be captured.

From a practical perspective, although the smaller the time step size the more theoretically "accurate" the simulation, industrial application of the CFD technique may not have such luxury of time. From the author's experience, to keep a balance between computational time and accuracy, selection of the time step size may be further influenced by the required number of iterations within a time step to reach convergence within each time step. Small time step size will lead to fewer iterations being required for each time step. However, small time steps will require more time steps to achieve the desired total simulation time. In contrast, relatively larger time steps will require more iterations within each time step to reach convergence because the changes in the flow field are larger. The larger time step will, however, lead to fewer time steps to achieve the total required simulation time. A delicate balance needs to be achieved to determine these parameters. Dynamic time stepping is certainly an option to "optimize" the required number of iterations for each time step. If the flow fluctuation is to be quantified with fast Fourier transform (FFT), which will be described later on, dynamic time stepping may result in inconsistent velocity sampling from the simulation, which may affect the analysis. This will be described in detail below.

As delineated earlier, another important consideration for choosing the time step size is that it affects the maximum sampling frequency of the flow field velocity. This may put a limit on the upper frequency in the analysis of the flow should an FFT type of analysis be undertaken. The FFT technique is a useful approach (Guo et al. 2001a,b) to analyze the long time scale and short time scale coherent fluctuation in the flow field of a spray dryer. In the flow field of a spray dryer, these flow field fluctuations with different time scales may occur simultaneously and they may overlap with each other. The FFT analysis will transform the time series of these fluctuating signals to the frequency spectrum so that the amplitude of the spectrum, delineating the dominance or significance of each amplitude, can be discerned. The frequency spectrum may be used to characterize the degree of airflow fluctuation within the spray chamber.

Chapter 4

Atomization and Particle Tracking

There are various types of atomizers used in spray drying and their details can be found in the *Industrial Drying Handbook* (Mujumdar 2014). For the discussion in this book, two broad types are classified: rotating atomizers and nozzle atomizers. Rotating atomizers encompasses atomizers, which utilize a rotating disc to atomize the droplets by imparting spin to the fluid. Some rotating atomizing plates utilize holes at the side of the rotating plate, whereas vanes are used in some designs. An important feature is that droplets are generated from the breakage of thin films of the feed material developed by the high-speed rotation.

The nozzle atomizers include pressure-based nozzles, one fluid nozzle or even two or more fluid nozzles. Some nozzles have internal swirl elements, which swirl the fluid upon exit. For the two fluid nozzles, the generation of the droplet (shearing of one of the fluid) may occur internally within the nozzle or externally from the nozzle. Although these different nozzles have inherently different mechanisms in generating the droplets, their implementation in the numerical framework is rather similar. Therefore, they are dealt together in Section 4.2.

Numerical implementation of the rotating atomizer is given in Section 4.1.

4.1 Capturing Rotating Atomization

Figure 4.1 shows an overhead snap shot illustration of how droplets are generated via a rotating disc atomizer. It can be seen that the rotation of the disc will "stretch" the liquid feed into a thin film from which tiny fine droplets will be formed by further instability of the film. The formation of the film is also determined by the shape or size of the vanes or holes. Owing to the random nature of the instability of the film formed, the size of the droplets generated will not be uniform. Therefore, we can expect a distribution of the atomized droplets. It can be visualized that due to the rotation of the film, as imparted by the rotating disc, the droplets will exhibit an initial velocity with velocity components tangential and radial to the rotating plate.

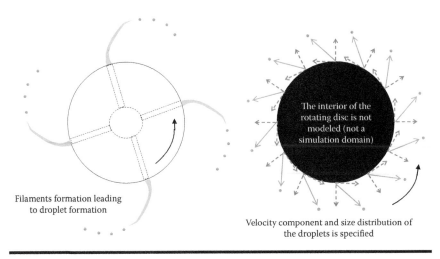

Filaments formation leading to droplet formation

The interior of the rotating disc is not modeled (not a simulation domain)

Velocity component and size distribution of the droplets is specified

Figure 4.1   Formation of droplets in a rotating atomizer versus the numerical representation (only four orifices are shown illustrated on the left for simplicity).

In a CFD simulation, the formation of the stretched film and the "budding off" of the tiny droplet is not of interest in a large-scale simulation of the spray dryer. The shape and size of the vanes or holes are also not important. Only the "end product" of the atomizer is important. Therefore, only the information on the droplet size distribution and the initial velocity or trajectory of the droplets will be of interest. With this in mind, the numerical implementation of the atomizer can be simplified, as shown in Figure 4.1. A few reports utilizing this approach are available in the literature (Huang et al. 2004; Woo et al. 2008a,b,c,d, 2012). Numerous droplet injection points at the periphery of the rotating disc can arbitrarily represent the formation of the droplets. In this approach, the higher the number of injection points, the more realistic the representation of the atomization process. As illustrated in Figure 4.1, this need not correspond to the number of holes or vanes of the rotating plate as, due to the rotation, the droplets will be generated along the whole periphery of the plate. However, the number of injection points has to be balanced with the possible instability, particularly if it is a transient simulation. From the author's experience, having a large "concentration" at the region adjacent to the atomizer will result in very large momentum, mass and heat exchange, which may lead to an unstable simulation. The total liquid feed flowrate is then divided equally between the injection points. At each injection point, a size distribution is then specified for the injected droplets. The method to obtain or to approximate the droplet size distribution is given in Section 8.1.1.

In the simulation domain, the rotating disc and the static housing are normally incorporated as part of the boundary of the simulation domain; the interior of the atomizer is typically not numerically captured in the simulation. This is shown in Figure 4.2. Numerically, it is very important to split the specified static housing and the rotating disc as separate walls. While the wall of the static housing is set as a static surface, the wall of the rotating disc (the bottom wall and the

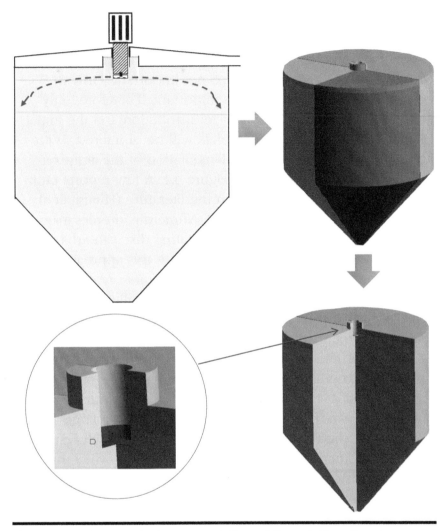

Figure 4.2 Geometrical representation of the rotating disc atomizer (surface in red is set to rotate and represents the side and bottom surfaces of the rotating plate).

side wall) is set as a rotating surface to impart a rotational momentum to the flow field.

The effect of the rotating wall on the flow field is very important and was experimentally and numerically verified to significantly affect the flow field (Langrish et al. 1992; Southwell and Langrish 2000; Woo et al. 2012). Specifically, the rotating wall affects the flow field in two ways. Firstly, it imparts a central rotating core in the region below the atomizer like a "tornado" which may extend towards the bottom region of the spray dryer. Experimental evidence showed that to a certain extent this imparts a higher stability to the air flow in the spray dryer (Southwell and Langrish 2000); minimizing the flapping motion of the central jet. However, in a separate simulation report for a short form spray dryer, the rotating central core imparted by the atomizer also induced significant swirl to the central airflow which splits recirculating airflow (Woo et al. 2009a,b). This affects the jet-feedback mechanism as discussed earlier in Section 3.2.

The influence of the pumping effect has been studied extensively (Langrish et al. 1992). The rotation of the disc and the flow of the liquid within the atomizer will tend to draw the air external to the atomizer into the gap between the rotating disc and the static part or housing of the atomizer. It was found that this significantly affected the stability and the flow field adjacent to the atomizer. From the author's experience in operating a rotating disc atomizer for a pilot-scale dryer (Woo et al. 2007a,b), such a pumping effect may recirculate the particles back into the gap, which may restrict the movement of the atomizer, leading to disruption in the process at extended operation times (atomizer plate jamming and high "amp" situation).

Some simulations are carried out to model the simulation of the rotating disc in great detail (Ullum 2006; Lin and Phan 2013). In these simulations, the airflow pattern and the pumping effect of the atomizer can be captured. However, most large-scale CFD simulations of the spray dryer do not account for the pumping effect. There was only one report in which detailed simulation of the rotating disc was incorporated into a large CFD simulation of a complete spray dryer (Ullum 2006). In the simulation, it was found that increasing the rotation of the atomizer significantly affected the simulation of the flow field. Although it was clear that such an effect is significantly affected by the rotating disc, it was unclear how much the pumping effect contributes to the overall flow field, particularly at regions away from the atomizer.

4.2 Capturing Pressure-Based Atomization

Figure 4.3 shows a snap shot of how droplets are generated via a pressure nozzle with internal swirl. This nozzle design was arbitrarily selected to illustrate the ideas included here which can be generalized to other nozzles. Similar to the rotating disc atomizer, the key phenomenon involved is to form thin films of the feed fluid such that tiny fine droplets can be budded off from the unstable swirling film. Due to the random nature of the instability of the film formed, the size of the droplets generated will not be uniform and we can expect a distribution of the atomized droplets. The velocity of the droplets generated will approximately follow the initial movement of the film. Therefore, if the film were initially swirled, there would be a tangential component to the initial velocity

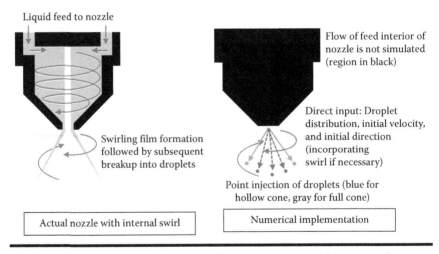

Figure 4.3 **Formation of droplets in a pressure nozzle versus the numerical representation (blue injection points—hollow cone and blue + gray injection points—full cone).**

of the droplets. Two important aspects which are specific to the nozzle atomizer are (a) the cone angle of the spray and (b) whether the spray of the cone is full or not. A full cone spray refers to a spray in which the whole cone is filled with droplets. In contrast, for a hollow spray (not full cone) only the peripheral of the cone has droplets forming.

In a CFD simulation, similar to the rotating atomizer, the formation of the stretched film and the "budding off" of the tiny droplet is not of interest in a large-scale simulation of the spray dryer. Only the information on the droplet size distribution and the initial velocity or trajectory of the droplets will be of interest. With this in mind, the numerical implementation of the atomizer can be simplified to that shown in Figure 4.3. In contrast to the rotating atomizer, the pressure atomizer is often represented by injection of droplets at a single point in space in the simulation. This single location is normally arbitrarily positioned just below the position where the orifice of the atomizer should be found. At that point in space, numerous droplet injection points are set overlapping each other.

While the velocity magnitude is normally set following the initial velocity of the droplets, the initial direction of the droplet trajectory is set to reflect on the cone angle of the spray. If a full cone spray is to be modeled, additional droplet injection points need to be set to account for the injection of droplets within the outer periphery of the cone. The total liquid feed flowrate is then divided equally between the injection points. Similarly, at each injection point, a size distribution is then specified for the injected droplets. The method to obtain or to approximate the droplet size distribution is given in Section 8.1.1. In most large-scale simulation of the spray dryer, the nozzle lance and any associated housing of the nozzle are typically set as a boundary of the simulation domain as the internal phenomenon within the nozzle or the lance does not directly affect the prediction of the flow field. Therefore, there is also no need to discriminate the orifice of the nozzle in the setting up of the model.

For nozzle-based atomizers, if it is a two-fluid nozzle, the air exiting the nozzle may affect the flow field adjacent below the nozzle. In most simulations reported in the literature *hitherto* this effect is commonly ignored. However, if desired, inlet airflow or a momentum source term may be easily incorporated into the simulation. In any case, the effect of the spray on the airflow region below the nozzle exit can be accounted for, in a simplistic manner, by the momentum transfer of the droplets injected into the simulation domain.

An important note: It may seem possible that with the use of the volume-of-fluid (VOF) technique, the film formation process of the atomized fluid can be further investigated and optimized in the nozzle design or rotating disc design. Although these are useful to understand and design the nozzle of interest, in the opinion of the author, such VOF simulations are not applicable for large-scale CFD simulation of the spray dryer or in the prediction of the atomized droplet size distribution. This is because the resolution of the mesh in a large-scale simulation will not be sufficiently fine to allow the VOF simulation to predict the formation of the droplet. Figure 4.4 shows a VOF simulation undertaken for an internal swirl nozzle. While the simulation can capture the formation of the swirl, it will not be sufficient to predict the formation of the individual droplets. This is because the length scale of the interior of the nozzle which is captured by the simulation is several magnitudes larger than that of the droplets; excessively high mesh refinements will be required to capture the formation of the fine droplets.

4.3 Simulating Particles Transport by Convection and Dispersion

Once a droplet is generated, it will then travel across the flow field or the domain of the spray dryer. The movement of the particle is affected by the drag exerted on the particle by the air and the gravitational force.

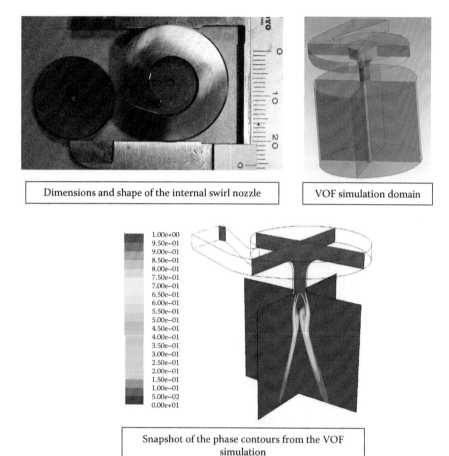

| Dimensions and shape of the internal swirl nozzle | VOF simulation domain |

Snapshot of the phase contours from the VOF
simulation

Figure 4.4 **VOF simulation of a pressure nozzle with internal swirl.
(With permission from Dairy Innovation Australia.)**

$$\frac{dv_p}{dt} = C_D \frac{18\mu_b Re}{24\rho_p d_p^2}(v_b - v_p) + \frac{g}{\rho_p}(\rho_p - \rho_b) \qquad (4.1)$$

Where the drag force coefficient C_D can be expressed by
the following for $0.5 < Re < 1000$,

$$C_D = \frac{24}{Re}(1 + 0.15Re^{0.687}) \qquad (4.2)$$

And the Reynolds number is computed based on the magnitude of the relative motion between the air and the particle, taking the diameter of the particle as the characteristic length,

$$Re = \frac{\rho_b d_p |v_p - v_b|}{\mu_b} \tag{4.3}$$

The force balance above is a simplistic form accounting for only the drag force of air and gravitational pull exerted onto the particle. The buoyancy force is also incorporated but in a particle–air system, owing to the low density of air, this effect will be very minimal. This form of the force balance is commonly used in the spray-drying simulations for the range of particles size and airflow velocity found in spray dryers. Additional forces such as particle rotation or thermophoretic forces can be easily added into the computation if required. Particles collisions are normally neglected owing to the low particle loading in the system, unless agglomeration or coalescence modeling is adopted.

There are numerous integration schemes that can be employed to track the position and velocity of the particles based on the force balance (e.g., the Runge Kutta approach), which will not be covered in detailed in this book. The airflow conditions throughout each integration are normally treated as constants, which reduces the integration into an ordinary differential equation. Depending on the dimension of the simulations, for the Cartesian framework, this integration is undertaken for the individual components of the particle velocity. By integration, the momentum (also heat and mass) source term for the airflow can then be computed. By repeatedly integrating this equation, the particles are "tracked" throughout the simulation domain.

One question, which is commonly asked, is on how the Eulerian–Lagrangian approach then captures the dispersion of the particles due to small-scale turbulence in the airflow. The following section is described following the author's familiarity in using the stochastic discrete random walk approach with the FLUENT framework. This aspect is particularly important, especially if the RANS-based turbulence model is employed in the airflow prediction as the smaller scale turbulence is only modeled and is not directly reflected by the velocity of the airflow predicted. One approach is by incorporating a random component in the local air velocity used in the computation of the drag force.

$$v_b = \bar{v}_b + v_b' \qquad (4.4)$$

The random "fluctuating" components can then be calculated corresponding to the turbulence parameters from the flow field calculation such as the turbulent kinetic energy and the dissipation length scale with the aid of a random number selection. Various forms of the calculation are available depending on the turbulence model used and will not be covered in detailed in this discussion. An important aspect to note is that a new "fluctuation" is calculated each time a particle is tracked across an integral time scale. Higher local turbulence will result in a larger integral time scale and vice versa. Such dispersion modeling was found to adequately describe the turbulent dispersion of water sprays (Nijdam et al. 2006). To the best of the author's knowledge, detailed dryer-wide validation of the dispersion modeling has yet to be reported in the literature.

Another question, which is commonly asked, is whether the Eulerian–Lagrangian approach utilized in spray-drying simulation is the same as discrete-element modeling (DEM). Both approaches are essentially the same in principle in tracking the movement of the particle throughout the simulation domain. The difference in mainly on how particle–particle and particle–wall collision is numerically captured. In spray-drying simulation, within the Eulerian–Lagrangian framework, owing to the low particle loading, the collisions mentioned above are normally treated as instantaneous, particularly if agglomeration or coalescence modeling is required. In contrast, the convention "DEM simulation" is normally commonly used to simulate systems with dense particle loading (e.g., fluidized bed or in particle pneumatic conveyancing). For such a system, as the particles are constantly in contact with each other, the collision force balance is not treated as instantaneous and changes throughout the integration of the force balance.

In an actual spray-drying process, there will be millions if not billions of droplets or particles in the spray chamber. It will be impossible to capture and compute the trajectory of all the droplets or particles. The Lagrangian–Eulerian approach typically used in most of the CFD simulations of spray dryers represents particles/droplets as parcels moving within the drying chamber. Each parcel then represents a collection of particles/droplets of the same size. It is also assumed that they will move in unison based on the assumption that particles/droplets of the same size and density interact in the same way as the local airflow conditions. Therefore, in calculating any momentum (heat and mass as well) transfer for a particular parcel, the total momentum exchange calculated has to be multiplied by the total number of particles represented by the parcel. In terms of heat and mass transfer, it is also assumed that all the particles in the parcel will experience the same heat and mass loss or gained. In other words, the number of particles represented by a parcel will remain constant throughout the simulation domain tracking. However, if agglomeration or coalescence modeling between the particles are incorporated, this will change the number of particles represented by a parcel.

4.4 Important Numerical Strategies in Two-Way Coupling

As discussed in Section 3.1.1, most spray-drying conditions will require full three-dimensional transient simulations, while there may be certain combinations of airflow rate and geometrical ratio leading to steady-state flows. When incorporating droplet injections into the flow field, it will be very important to start with a fully developed flow field without particle injection; that is, the "dry run." The strategies to obtain the fully developed "dry run" flow field are described in Section 3.1.2. The following section was prepared mainly based on the familiarity and experience of the author with the FLUENT package. However, the ideas presented can be extended to various platforms.

In steady-state simulations, individual particles are injected into the developed flow field and tracked through the developed flow field with only the flow field affecting the trajectory of the particles and not vice versa. As the particle is tracked through the flow field, only the heat, mass and momentum of the particle are updated. Any corresponding change to the flow field owing to the property exchanges will be stored as source terms in the numerical framework. After all the particles are tracked, the flow field is then recalculated incorporating the source terms from the particle tracking until the local residuals are sufficiently low and acceptable. This numerical scheme is then repeated iteratively until the solution converges and via this approach two-way coupling is achieved. Convergence can be determined by monitoring the changes in the outlet airflow humidity and temperature; convergence is achieved when there is minimal change at subsequent iterations. Typically, there will be minimal change in the momentum of the airflow at the outlet even with the incorporation of droplet injection. This iterative scheme is shown in Figure 4.5.

For steady-state droplet injection, each droplet tracking is undertaken independent of each other. Therefore, this will be less computationally intensive as only one particle is tracked at a time. In view of this, there is a possibility to significantly increase the number of droplets injected into the simulation domain. As mentioned earlier, the more particles injected into the domain, the more realistic is the simulation. However, from the author's experience in a steady-state simulation of

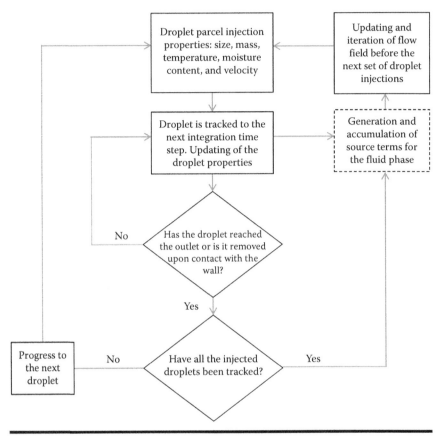

Figure 4.5   Two-way tracking numerical scheme for the steady-state particle injection.

a spray dryer, there will be a threshold in the number of particle injections, above which there will be insignificant changes in the predictive capability of the simulations. It will be important for the modeler to evaluate this for their simulations.

In transient simulations, this iterative scheme is repeated within each numerical time step. However, at each time step, the droplets or particles are tracked to move within the time step and not throughout the entire domain until they exit. Convergence of the simulation is then determined by tracking the changes of the outlet air humidity and temperature as a function of the simulation time and not between iterations within a time step. From the author's experience, monitoring both parameters is very important as the change in temperature of the airflow may stabilize significantly faster than the change in the outlet air humidity. It will be important to ensure that both parameters are stable to avoid any false indication of a fully developed flow field with particle injection. What is not shown in Figure 4.6 is that there will be discrete injection of droplets into the simulation domain, which may or may not follow the time step of the fluid flow. This will progressively increase the number of tracked particles in the simulation domain until an apparent "steady" value is reached when particles have travelled towards and start leaving the simulation domain. It should be noted that even if the number of particle built up in the simulation domain has reached an apparent "steady" value, it is not an indication of a fully developed flow field.

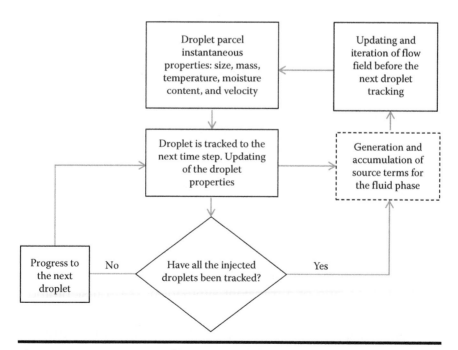

Figure 4.6   Two-way tracking numerical scheme for the transient particle injection.

Two considerations below are important for transient simulations:

1. The total number of particles in the simulation domain in a developed "pseudo-steady" flow: This translates to the number of particle trackings that need to be undertaken at each time step. Depending on the computational resources available, too many particles will lead to very long computational time in a transient simulation. The total number of particles will depend on the number of particle injected and the frequency of injection. Consider the following simplistic example in which the average residence time of the particles in the spray dryer is 6 s. If ten droplets are injected at an interval of 0.01 s, there will roughly be 6000 particles in the simulation domain at any one time at the pseudo-steady state. The number of particle injections is typically decided based on the need to capture the size distribution and geometrical effect of the atomizer (for the rotating disc). The author typically uses injection intervals following the airflow time step intervals or at multiples of the airflow step intervals. This is mainly for consistency so that the particle injection and the airflow phenomenon are captured on a similar time scale. Similarly, the author preferred to track or progress the particle trajectory once per airflow time step. Ideally, it will be useful to vary these parameters to a threshold from which the simulation will be independent of the parameters. However, as the transient simulation is relatively long, the need for such a detailed sensitivity test may need to be further evaluated depending on the resources and time available for a project.
2. Postprocessing of a transient simulation: For transient and steady simulations, once the flow field with droplet injection is deemed fully developed, it is important to ensure that the overall mass and energy balance at the inlet and outlet of the simulations are balanced to the acceptable

threshold. Interpretation of the simulations should only be undertaken with the developed flow field.

For the steady-state simulations, analysis on the particle trajectories can be undertaken by simply injecting and tracking any individual particles in the flow field incorporating fully developed with particle injection source terms; two-way coupling was adopted in the development of the flow field. This form of analysis assumes that the particle injected for the postprocessing analysis is representative of the particle trajectories that would have affected and caused the formation of the fully developed flow field. Analysis for a steady-state simulation is relatively fast.

For transient simulations, the conventional approach for analysis is to collect statistical data of the particles at extended simulation time beyond the time required to achieve the fully developed flow field incorporating two-way coupling. As an alternative to this time-consuming analysis method, the author has explored the use of a pseudo-steady analysis method. The key assumption in this form of analysis is that once fully developed self-sustained oscillation is achieved in the fully developed flow field, the flow field at any instance of the transient analysis is representative of the average behavior of the spray dryer. With this assumption, a single flow field can then be used to analyze the particle trajectory using a steady-state approach. This approach of analysis was adopted successfully to analyze a transient simulation of a low-velocity spray tower (Woo et al. 2011b). This approach so far has only been evaluated for spray-drying chambers with "axis-symmetric" geometrical features. It is unclear how this approach can be suitably applied for nonaxis-symmetric spray dryers; this is particularly for spray dryer with nonbalanced position of the air outlets or with nonbalanced outlet air protrusions.

Chapter 5

Droplet Drying and Quality Modeling

5.1 How Is Drying Captured in the Particle-in-Cell Approach?

As discussed in the preceding chapter, in the Lagrangian–Eulerian approach a particle is numerically tracked throughout the simulation. Throughout the particle's trajectory, it exchanges momentum with the local airflow. For a full CFD simulation, two-way coupling is normally adopted in which the airflow exerts a certain amount of drag on the particles, pushing the particle; at the same time, the drag effect also changes the momentum of the particle. A similar approach is adopted to numerically capture the drying process of the particle within the simulation domain. As the particle or droplet moves within the chamber, heat and moisture are exchanged with the local airflow contacting the particle. This is illustrated in Figure 5.1.

Heat and mass transfer during the drying process can be mathematically represented by the following expressions, adopting the interface film transfer approach. Other mathematical forms may be adopted and the following equations are

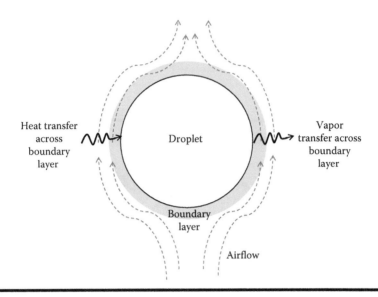

Figure 5.1   Heat and mass transfer exchange between the droplet/ particle and the local airflow flow (although not shown for simplicity, the thermal and boundary layers will have different thicknesses).

adopted mainly due to the preference and familiarity of the author to express some of the ideas in this chapter.

$$-\frac{dm_w}{dt} = h_m A_p (\psi \rho_{v,surface} - \rho_{v,bulk}) \tag{5.1}$$

$$\frac{dT_p}{dt} = \frac{hA_p(T_b - T_p) - (dm_w/dt)\Delta H_L}{m_p C_{p,p}} \tag{5.2}$$

Mass transfer can transfer into or out of the droplet depending on the humidity of the local airflow. In the energy transfer equation, removal of moisture by drying always removes energy from the droplet, whereas convective heat transfer may be into or away from the droplet depending on the temperature difference between the local air and the droplet. Local airflow here refers to the air at the control volume at the instance where the droplet is located within each particular time step. At this point, it will be important for the reader to be familiar with the discretization concept as described in Chapter 2 in order to make full use of the following discussion. Therefore, within each time step, the mass and temperature of the droplet are updated as follows:

$$m_{w,new} = m_{w,old} - \Delta t h_m A_p (\psi \rho_{v,surface} - \rho_{v,bulk}) \qquad (5.3)$$

$$T_{p,new} = T_{p,old} + \Delta t \frac{h A_p (T_b - T_p) - (dm_w/dt) \Delta H_L}{m_p C_{p,p}} \qquad (5.4)$$

Conversely, the mass of the evaporated water and the energy in the local airflow can be updated. It can be deduced that the terms with the time step in Equations 5.3 and 5.4 represent the source terms for the overall mass and energy balance for the local airflow. A constant source term is normally assumed within a single time step depending on the solution framework. Depending on the CFD package used, updating of the evaporated moisture in the control volume may require incorporating another species transport equation into the simulation. Such an approach will conveniently allow computation of the transport of evaporated moisture in the simulation domain, allowing the humidity of the air to be tracked. While the updating on the air phase may not pose a problem, a few important aspects have to be considered so that the numerical updating is physically realistic for the particles. A few physical constraints apply

1. Should there be a positive temperature increase in the particle (the source term), the updated particle temperature can never be higher than the temperature of the local airflow. This is because of the temperature driving force requirements for convective heat transfer into the droplet, which limit the maximum temperature of the droplet. One may argue that condensation may increase the temperature of the particle beyond the air temperature,

but this may not be as significant as the convective heat transfer in an actual spray-drying operation.

2. The updated particle moisture can never be lower than the equilibrium moisture corresponding to the humidity and temperature conditions of the local airflow. This is a fundamental aspect of drying.

3. It can be noted that the two numerical constraints above arises mainly due to the Euler form of the heat and mass transfer equations. A higher order discretization may minimize but may not completely alleviate these potential numerical discrepancies.

In the numerical treatment above, the updating of the source term is mainly determined by the time step size. Therefore, when implementing droplet drying models in a CFD simulation, very small integration time steps may be required; this is a parameter that has to be estimated by the modeler. It may also be a good numerical practice to include a numerical limiter in the droplet drying code. Such a limiter may not be available in commercially available codes as some codes are designed for pure water evaporation, in which case, the second physical constraint does not apply. Limiting the lower bound of the particle moisture may require knowledge of the sorption or desorption isotherm of the material.

When arbitrarily limiting the updated particle temperature and moisture, it may also be useful to limit the updating of the continuous phase. When using commercial packages, one has to check if the source terms calculated are directly returned to the solver. This arbitrary limiting of the particle updating does not modify the calculated source term to reflect on the numerical limiter put in place. Nevertheless, if the time step size is sufficiently small, air source term-particle updating discrepancies introduced by the numerical limiter may be minimized.

It should be noted that the sorption isotherm of a material may vary, depending on the state of the material. Take lactose for example. Monohydrate crystallized lactose may have minimal moisture content apart from the chemically bound water. Spray-dried lactose particles on the other hand may be amorphous in structure and will exhibit higher hygroscopic behavior and will have higher equilibrium moisture content. Therefore, when selecting suitable sorption isotherm data for implementation into the CFD model, it will be crucial to mea-sure them from samples closest to that expected from spray drying. This will be an important consideration when explor-ing new spray dry formulations.

It will be evident by now that in Equation 5.1 there is a sur-face relative humidity term that augments the droplet surface vapor pressure which drives the evaporation. If the droplet remains as a pure droplet (evaporation of water, etc.) the sur-face relative humidity term will then remain at unity. However, if the droplet initially contains dissolved solids, as moisture is being removed and the concentration of the solute increases, the relative humidity at the surface will drop. This is due to the evaporation retardation of the solutes. Towards the latter stages of drying, the solutes may even solidify, further retard-ing the evaporation process. Such a drying retardation process then further manifests in the energy transfer equation source term.

The bulk of droplet drying research available in the litera-ture, in fact in many reports on drying in general, pertains to understanding and quantifying this evaporation retarda-tion process. A droplet drying model refers to the mathemati-cal framework used to quantify this evaporation retardation behavior. In the following section, a few well-established drop-let drying models suitable for implementation in large-scale CFD simulations are introduced.

5.2 Comparison on the Existing Models

To the best of the author's knowledge, the majority of CFD simulations for spray dryers utilize the lump droplet drying approach in which the particle moisture distribution across the radius of the particle is not considered. Equations 5.1 and 5.2, introduced earlier in this chapter, are an illustration of a lump droplet drying approach where the mass and energy transfer at the interface is of interest, not the transfer processes within the particle. Several droplet drying modeling approaches are well established, such as the reaction engineering approach (REA) and the characteristic drying curve approach, which can be found in the literature (Langrish and Kockel 2001; Chen 2008). These two lump approaches to drying provide a numerically inexpensive method for implementation in larger scale CFD analysis. The distributed diffusion-based drying approach, which discretizes the droplets into multilayers of shells, is more numerically intensive for implementation in CFD analysis (Adhikari et al. 2004; Handscomb et al. 2009).

It is very important to highlight this aspect of droplet drying modeling. The lump droplet drying approach assumes that the temperature profile and the moisture profiles within the droplet (across the radius of the particle) are relatively uniform. This is certainly a very fair assumption and an easily justified one for the temperature distribution. A Biot number analysis will reveal that within the range of convective heat transfer typically encountered in spray drying, relative to the thermal conductivity in most spray dried particles, temperature gradient within the particles may be neglected. A more accurate analysis using a modified Chen–Biot number incorporating evaporation in the analysis provided more support for this assumption (Patel and Chen 2008).

Assumption on the uniform moisture profile is mainly a simplifying assumption. A dehydrating droplet may experience significantly less moisture on the surface when compared to the internal region. Such a phenomenon may be amplified if the dissolved solids have skin-forming or crust-forming characteristics, which will tend to impede transport of moisture within the droplet, leading to depletion of moisture on the surface of the semi-dried solid or particle. Therefore, the lump droplet drying approach only provides the average particle moisture content. From an industrial point of view, this may already be sufficient as the product quality is mainly characterized with the average moisture content. Numerically, capturing the surface moisture may be important in the prediction of agglomeration or prediction of deposition on the wall of the chamber, which is in essence a particle surface phenomenon. This will be covered in Chapter 6 on modeling agglomeration.

How the two types of lump droplet drying model are implemented in a CFD simulation framework is discussed below.

5.2.1 Characteristic Drying Curve

The premise of the characteristic curve drying is that the behavior of the falling rate period of a certain material under convective dehydration is similar, regardless of the size of the droplet or the external ambient drying conditions. Mathematically, to capture this unique unified drying behavior (Langrish and Kockel 2001), the CDC adopts two sets reference "points" which are (1) the evaporation rate at the initial constant rate period as a reference to denote the reduction in drying rate and (2) the critical moisture and the equilibrium moisture to denote the extent of dehydration.

Usage of the evaporation rate at the initial constant rate period has important physical significance as this is the maximum drying rate that can be achieved corresponding to the ambient thermal and humidity conditions. Therefore, at the initial instance of the falling rate period, the ratio of the actual drying rate at that instance to the drying rate, at the constant rate period, will be unity. As the drying rate drops when the droplet progresses into the falling rate period, this ratio then drops and eventually reaches zero when evaporation ceases. Therefore, this provides a very convenient way to capture the reduction of the drying process. In the constant rate period, it is assumed that the drying rate is controlled by the rate of heat transfer into the system,

$$-\frac{dm_w}{dt} = \frac{hA_p(T_b - T_{p,wb})}{\Delta H_L} \tag{5.5}$$

How do we then denote the extent of drying in reducing the relative drying rate given by Equation 5.7? The critical moisture content denotes the "dryness" or "wetness" of the droplet corresponding to the initiation of the falling rate period. Once the droplet enters the falling rate period, dehydration will only cease once the droplet reaches the equilibrium moisture corresponding to the external ambient temperature and humidity conditions. Therefore, the equilibrium moisture of the particle denotes the extent of drying corresponding to the end of dehydration. Therefore, the positive difference between the critical moisture and the droplet moisture denotes the extent of drying into the falling rate period. Expressing this as a ratio to the maximum extent of drying in the falling rate period conveniently gives a dimensionless parameter with a value of unity to denote the extent of drying,

$$f = \left[\frac{X - X_{eq}}{X_{er} - X_{eq}}\right]^N \tag{5.6}$$

Combining both equations, the premise of the CDC approach to describe the falling rate drying behavior then becomes,

$$-\frac{dm_w}{dt} = \left[\frac{X - X_{eq}}{X_{cr} - X_{eq}}\right]^N \frac{hA_p(T_b - T_{p,wb})}{\Delta H_L} \tag{5.7}$$

The moisture ratio is conveniently used to reduce the drying rate relative to the maximum drying rate of the droplet corresponding to the initial constant rate period. A "n" parameter is then introduced to account for the material-specific behavior. Determination of this material-specific factor has to be undertaken experimentally. The experimental approach will be discussed in more detailed later on. For the industrial reader, when communicating with the CFD modeler, this parameter is commonly referred to as the "drying kinetics" in the CDC framework. This is because this parameter determines the rate or behavior of the dehydration process. A large part of a CFD simulation project may actually pertain to experimentally obtaining these parameters. There are a number of experiments reported in the literature describing this parameter for some common spray-dried materials. These are given in Table 5.1.

In Table 5.1, it can be seen that the critical moisture content is specified. This is because it is another material-specific parameter, which has to be obtained experimentally. It is important to note that the critical moisture content, on top of being material specific, may also be affected by the initial concentration of the feed solution to be spray dried and the ambient drying conditions (Zbicinski and Li 2006). The critical moisture content denotes the extent of drying where the surface of the droplet starts to impede free moisture evaporation. This impeding phenomenon may be due to the increase in the solute concentration at the surface or may even denote the initiation of crust formation or solidification at the surface. It is important to bear in mind the significance of the nonuniformity in the moisture content across the radius of the droplet as discussed earlier. A higher initial solute concentration will tend to lead to earlier solidification or solute concentration increase on the surface, translating to higher critical moisture concentration. Similarly, if the ambient condition provides a higher potential for dehydration, a faster solidification of the droplet surface may be expected.

How do we then experimentally determine these parameters for new formulations? These can be experimentally

Table 5.1 Characteristic Drying Curve Parameters for Some Common Spray-Dried Materials

Materials	References	CDC Index, -	Critical Moisture Content
Skim milk	Langrish and Kockel (2001)	1 (linear)	Initial moisture content (80% wt was analyzed)
Sucrose	Woo et al. (2008a,b,c,d)	2.58	Initial moisture content (50% wt analyzed)
Maltodextrin	Woo et al. (2008a,b,c,d)	3.22	Initial moisture content (50% wt analyzed)
Sucrose-maltodextrin (1:1)	Woo et al. (2008a,b,c,d)	1.98	Initial moisture content (60% wt analyzed)
Maltodextrin	Zbicinski and Li (2006)	1 (linear)	From pilot-scale conditions: Evaporation rate >8 g/cm² s (10% wt initial moisture) Evaporation rate >5 g/cm² s (30% wt initial moisture) Evaporation rate >3 g/cm² s (50% wt initial moisture) The critical moisture content was close to the initial moisture

determined by dehydrating a sample of the material of interest under convective hot air conditions and recording how the mass of the sample changes over the dehydration time. A sample mass against the dehydration time is firstly obtained at different drying conditions by varying the drying temperature (and if possible the humidity and velocity of the convective hot air). The data are then translated into the form of "rate of dehydration" against the moisture content of the droplet. From the translated data, the critical moisture content is then determined by examining the moisture content in which the dehydration rate starts to drop. The next step is then to normalize the data with the theoretical reference points discussed above and collapse the data to find the "n" parameter by curve fitting the falling rate period. This process is summarized graphically in Figure 5.2.

5.2.2 Reaction Engineering Approach

The REA visualizes the dehydration process as one that is governed by an activation energy, akin to a chemical reaction (Chen 2008; Chen and Putranto 2013). In essence, for dehydration to occur there is an activation energy which the dehydrating system has to overcome. As moisture is progressively lost from the system, this activation energy progressively increases, thus making the dehydration more difficult and retarding the dehydration rate. This approach does not clearly distinguish a constant rate period or a falling rate period but is mainly based on a continuous change or increase in the activation energy of the dehydration process. How is this mathematically captured? The REA approach utilizes the vapor concentration difference form in describing the dehydration process (Equation 5.1). The surface relative humidity term is then visualized and described by the Arrhenius form,

$$\psi = \exp\left(-\frac{E_v}{RT_p}\right) \qquad (5.8)$$

Combining this expression for the surface relative humidity term with Equation 5.1, the drying process can then be viewed as an activation energy process. The key element of the REA

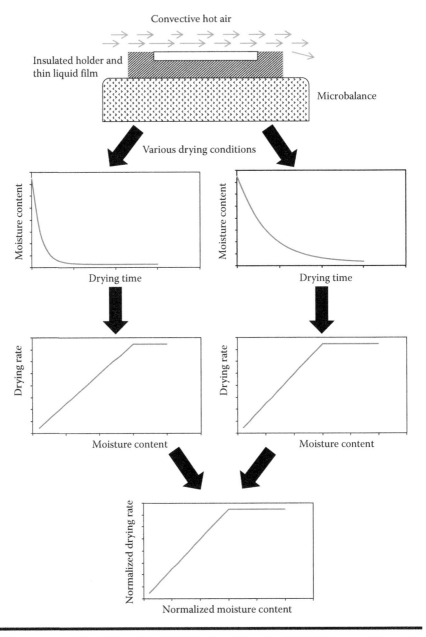

Figure 5.2 Obtaining the CDC drying kinetics (thin film experiment is shown as an example: if using single droplet, the changes in area need to be measured). Linear falling rate is shown as an example.

is in determining how the activation energy increases with the reduction in droplet moisture. Rearranging the equation, the activation energy can then be expressed as follows,

$$E_v = -RT_p \ln\left(\frac{-(dm_w/dt)/b_m A_p + \rho_{v,bulk}}{\rho_{v,surface}}\right) \qquad (5.9)$$

This equation shows that the activation energy of the drying process can then be obtained experimentally by finding the following parameters at any particular instance in dehydrating a droplet: (1) droplet diameter, (2) droplet rate of dehydration, and (3) droplet temperature. The mass transfer coefficient can be computed from well-established Sherwood correlations, whereas the surface vapor concentration can be calculated with the available empirical correlations, provided that the droplet temperature is known. Alternatively, the ideal gas equation can be used in combination with the Antoine equation for calculating the saturated vapor concentrations;, assuming ideal gas behavior. From the author's experience, within the drying condition ranges evaluated so far, both approaches will lead to rather similar predictions.

These parameters can be conveniently measured by dehydrating a single droplet; this will be described in detailed later on. By varying the drying air temperature, different sets of the activation energy versus drying time can be obtained. These can then be translated to the form of activation energy versus moisture content by calculating the droplet moisture content at each instance of drying. It will be observed that the activation energy is zero at the initial instance of drying and increases exponentially as the moisture content reduces. This is called the activation energy curve. The next step is to normalize the activation energy with the maximum activation energy corresponding to the external drying conditions used in the experiments, calculated as follows:

$$E_{v,\max} = -RT_b \ln\left(\frac{\rho_{v,bulk}}{\rho_{v,surface}}\right) \qquad (5.10)$$

This form is obtained by simplifying Equation 5.9 by assuming zero evaporation, which corresponds to the condition with the highest activation energy (Chen and Putranto 2013). This typically corresponds to the end of the drying process in which the particle also approaches the ambient air temperature. It should be noted that the particle surface vapor concentration is a saturated one, whereas that of the ambient bulk air has to include the relative humidity of the ambient air.

The moisture content is then normalized with equilibrium moisture of the particle corresponding to the drying conditions. It will be observed that all the normalized activation energy curves can be reduced and can collapse into a single curve. This is the material-specific master activation energy curve. For the industrial reader, obtaining the drying kinetics for the REA refers to the measurement and computation of this master activation energy curve. This whole process of obtaining the REA kinetics is graphically illustrated in Figure 5.3.

The master activation energy curve is the unique feature of the REA. It appears that visualization of the dehydration process as an activation energy process allows a unified approach in capturing behavior of a material. This means that once a single master activation energy curve is obtained, it can then be used to predict the activation energy of the dehydration process (delineating the progressive retardation of drying) corresponding to different local external drying conditions. This is done by reversing the process shown in Figure 5.3 and by substituting different maximum activation energy and

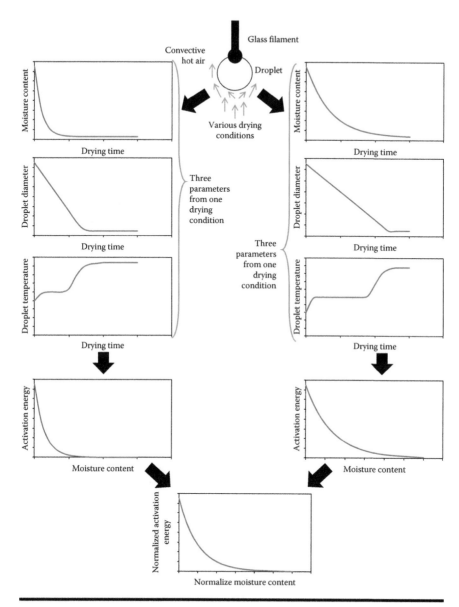

Figure 5.3   Obtaining the REA drying kinetics (single droplet experiment is shown as an example: if using thin film, the changes in area need not be measured).

equilibrium particle moisture corresponding to the external drying conditions. This approach has also been shown to effectively capture other forms of solids or bulk material drying (Chen and Putranto 2013). Table 5.2 shows a list of master activation energy curves reported in the literature for some commonly spray-dried materials.

Although the master activation energy can be used to account for different external drying condition, it is specific to the initial concentration of the solute in the droplet. This is because different initial solute concentrations will give different potential in the solidification of the droplet and this will lead to different drying histories of the droplet. Therefore, different master activation energy curves will have to be generated to correspond to various initial feed concentrations. This will be an important aspect to consider, particularly due to commercial spray drying interest to increase the feed concentration for better drying economics. An example is illustrated in Figure 5.4, plotting the equations for skim milk provided in Table 5.2.

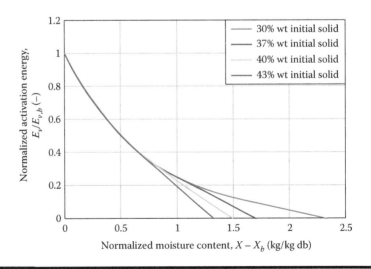

Figure 5.4 The effect of initial solute concentration on the REA kinetics. (Reproduced based on the equations by Chew, J.H. et al., *Dairy Science & Technology* 2013, 93, 415–430, listed in Table 5.2.)

Table 5.2 Master Activation Energy Curves for Some Common Spray-Dried Materials

Materials	References	Initial Moisture Content, % wt (Wet Basis)	Master Activation Energy Curve
Skim milk	Chew et al. (2013)	Various	Single normalized activation energy at normalized moisture content lower than the critical moisture: $$\frac{\Delta E_v}{\Delta E_{v,max}} = 1 - 1.305(X - X_b) + 0.7097(X - X_b)^2$$ $$- 0.1721(X - X_b)^3 + 0.0151(X - X_b)^4$$ Various normalized activation energy at moisture content higher than the critical moisture: (30% wt initial moisture − $X_{critical}$ 1.362 kg/kg db) $$\frac{\Delta E_v}{\Delta E_{v,max}} = -0.1617(X - X_b) + 0.3768$$ (37% wt initial moisture − $X_{critical}$ 0.969 kg/kg db) $$\frac{\Delta E_v}{\Delta E_{v,max}} = -0.3595(X - X_b) + 0.6129$$

(Continued)

Table 5.2 (*Continued*)   Master Activation Energy Curves for Some Common Spray-Dried Materials

Materials	References	Initial Moisture Content, % wt (Wet Basis)	Master Activation Energy Curve
			(40% wt initial moisture – $X_{critical}$ 0.788 kg/kg db) $$\frac{\Delta E_v}{\Delta E_{v,max}} = -0.4779(X - X_b) + 0.7129$$ (43% wt initial moisture – $X_{critical}$ 0.640 kg/kg db) $$\frac{\Delta E_v}{\Delta E_{v,max}} = -0.5927(X - X_b) + 0.7871$$
Full milk	Chen and Lin (2005)	20% wt	$$\frac{\Delta E_v}{\Delta E_{v,max}} = 0.957\exp\left(-1.291(X - X_b)^{0.934}\right)$$
WPC	Lin and Chen (2007)	30% wt	$$\frac{\Delta E_v}{\Delta E_{v,max}} = 1.335 - 0.3669\exp\left((X - X_b)^{0.3011}\right)$$
Cream	Lin and Chen (2007)	30% wt	$$\frac{\Delta E_v}{\Delta E_{v,max}} = 1 - 0.6282(X - X_b)^{0.5561}$$
Lactose	Lin and Chen (2006)	20% wt	$$\frac{\Delta E_v}{\Delta E_{v,max}} = 1.017\exp\left(-1.678(X - X_b)^{1.018}\right)$$

(*Continued*)

Table 5.2 (Continued) Master Activation Energy Curves for Some Common Spray-Dried Materials

Materials	References	Initial Moisture Content, % wt (Wet Basis)	Master Activation Energy Curve
	Fu et al. (2011)	10% wt	$\dfrac{\Delta E_v}{\Delta E_{v,max}} = 0.802\exp\big(-1.98(X - X_b)\big) + 0.198\exp\big(-(X - X_b)^{0.475}\big)$
Maltodextrin (DE6)	Patel et al. (2009)	50% wt	$\dfrac{\Delta E_v}{\Delta E_{v,max}} = \big(1 - 2.060(X - X_b)^{0.3593}\big)$
		40% wt	$\dfrac{\Delta E_v}{\Delta E_{v,max}} = \big(1 - 1.777(X - X_b)^{0.05746}\big)$
Sucrose	Patel et al. (2009)	50% wt	$\dfrac{\Delta E_v}{\Delta E_{v,max}} = \big(1 - 0.9438(X - X_b)^{8.8240}\big)\exp\big(-0.6030(X - X_b)^{2.0240}\big)$
		40% wt	$\dfrac{\Delta E_v}{\Delta E_{v,max}} = \big(1 - 0.03447(X - X_b)^{8.2950}\big)\exp\big(-0.5353(X - X_b)^{1.6860}\big)$
Sucrose-Maltodextrin (1:1)	Woo et al. (2008a,b,c,d)	60% wt	$\dfrac{\Delta E_v}{\Delta E_{v,max}} = \exp\big(-0.892(X - X_b)^{2.022}\big)$

5.2.3 Comparison between the CDC and the REA Approach

The discussion in this section is mainly derived from the experience of the author in comparing the two models published in a series of two journal articles (Woo et al. 2008a,b). One main difference between the models is in the way the effect of the initial moisture on the drying behavior is numerically captured. Different initial moisture contents affects the drying duration in which the inhibition of evaporation occurs. For a droplet with higher initial moisture content, any possible crust formation or dehydration inhibiting solid formation may only occur after significant moisture is loss, giving a lower critical moisture content. On the other hand, when a droplet has a lower initial moisture content, due to the presence of the high solute concentration, the inhibition of dehydration may occur immediately during drying, giving a high critical moisture content.

The REA accounts for the effect of different initial moisture content by utilizing a different activation energy master curve for different initial moisture contents. This aspect is very important as it significantly affects the initiation of moisture evaporation inhibition of the particle during the Lagrangian particle tracking. This numerical requirement means that interpolation between different activation energy curves is required when changing the initial moisture content of the droplet in the simulation. It should be noted that the REA approach does not assume any critical moisture for the initiation of dehydration inhibition; rather, it provides a "continuous" prediction of dehydration and the inhibition to dehydration is computed as a combination of the kinetics and the conditions of the particles.

On the other hand, the CDC approach accounts for the effect of different initial moisture content by variation in the critical moisture content. Derived from thin film isothermal experiments, there are suggestions in the literature to assume the initial moisture content as the critical moisture content. However, evaluation by the author showed that such an assumption might produce a physically nonrealistic drying history. In contrast to the REA mathematical formulation, which directly calculates the moisture inhibition characteristics based on the conditions of the particle as well as the ambient conditions, the evaporation driving force for CDC is only dependent on the external drying conditions. Therefore, empirical correlations may be required to correlate the critical moisture of the material to different initial moisture content as well the other parameters of the droplet and the ambient conditions. To the best of the author's knowledge, only one report in the literature illustrates how the critical moisture of maltodextrin, dehydrated from an initial liquid state, changes with different ambient drying conditions (Zbicinski and Li 2006).

Another feature of the REA approach is its potential to compute particle surface moisture using the lump drying approach (Woo et al. 2007a,b; Chen 2008). Conventionally, particle surface moisture prediction is only possible when a distributed moisture evaporation model is adopted. The distributed moisture approach may have very high numerical requirements, making it difficult to be implemented into a large CFD framework. The surface moisture computation is possible because the REA framework is based on the particle boundary layer vapor mass transfer as the driving force for evaporation (Equation 5.1); in contrast, the CDC framework is based on energy transfer as the driving force for evaporation. A few key assumptions required when using the REA for particle surface moisture calculation are (1) uniform temperature distribution radially within the particles and (2) the particle surface moisture condition is in an equilibrium state. The second assumption implies that the fractionality term calculated in the REA framework can be used in conjunction with the material isotherm to predict the particle surface moisture. This concept at the moment still requires refinement and validation with detailed experimental measurements.

In an industrial application, when the main aim is to predict the moisture content of the particles at the outlet, such differences between the two models may be very small, although the two models will produce different drying histories. A report published by the author indicated that the difference between the two models may be as little at 1%–3% wt moisture. This is mainly because of the rapid evaporation of the particle or droplet relative to the size of the chamber. The quantitative comparison provided above was derived from a 2 m high pilot-scale cylinder-on-cone spray dryer. For larger dryers, the may be longer residence time to reach approximately the equilibrium moisture of the material. Therefore, for the prediction of the particle outlet moisture content, the differences may be minimal. However, the differences in the drying history, particularly in the initial droplet–air contact, may result in contrastingly different particle quality prediction such as in-situ crystallization or denaturation of protein, etc. Evaluation of such differences on the particle stickiness had been reported by the author and it has significantly affected the prediction of particle deposition.

In addition to the numerical constraints provided in Section 5.1, implementation of the REA framework may require special attention due to the Arrhenius exponential terms (and possibly in the activation energy master curves) used in the equations. If at any point during the CFD simulation these terms become negative in magnitude, the simulation will crash. Logically, based on physical considerations, these terms should not become negative. From the author's experience in implementing these models in commercial codes such as FLUENT, however, the background "numerical crunching" may sometimes lead to such an occurrence. Therefore, it is important to put in numerical limiters for these terms to avoid simulation crash. Such numerical intricacies are not encountered in the CDC framework as there are typically no exponential terms involved.

5.3 How Is the Drying Kinetics Measured for Specific Products?

A key message from Section 5.2 is that the model and the kinetics used to capture the drying process in a CFD simulation framework should be able to reflect the changing local ambient drying conditions experienced by the moving droplet within the simulation domain. This is reflected by the "unified" theoretical approach as shown by the CDC and the REA. Therefore, the experiments undertaken to determine such kinetics has to be undertaken on the fundamental droplet-scale and dryer-wide measurements will not be useful in this endeavor. Although this requirement may be clearly understood by the modeler, industrial partners may have a different notion and may assume and further suggests the measurements of dryer-wide parameters to determine the drying kinetics. This possible misinformation should be clearly managed in the development of a CFD project. From the theoretical description of the models, regardless of whether CDC or REA, four parameters will need to be measured against the drying time: (1) mass of the sample, (2) temperature of the sample, (3) diameter or shrinkage of the droplet, and (4) equilibrium moisture content.

Thin film drying has been widely used to measure the drying kinetics of a material. In this approach, the initial liquid is thinly spread into a thin film and is exposed to convective hot air at controlled humidity, temperature, and velocity. As the film dries, the mass is continuously recorded. The same experiments can be further repeated under the same drying conditions with a small thermocouple dipped into the film to measure the temperature of the film throughout the drying process. This method, however, does not allow droplet shrinkage to be measured as it essentially does not capture the spherical geometrical aspect or a dehydrating droplet and only captures the drying behavior of the material. It is unclear at the moment whether the film shrinkage typically observed can be translated to actual droplet shrinkage or not. Figure 5.2 illustrates a typical thin film experimental set-up.

This approach is relatively simple and low-cost one to set up. A few important aspects need to be considered when using this approach to achieve a closer representation of the spray-drying process. Firstly, the film should be spread as thin as possible so that the internal moisture diffusion phenomenon can be reproduced as close as possible to a minute sprayed droplet. Even for very dilute low viscosity fluid, due to surface tension, a thin film may be difficult to achieve. In some reports, the surface in which the film is to be spread is roughened to help in the "holding" of the stretched film (Vigh et al. 2008). This spreading process may be difficult for feed solutions with high initial solute content and if it is viscous. In such cases, thicker film may have to be used and it is left to the judgement of the modeler as to whether the kinetics obtained is reliable or not. Another aspect is that the bottom region of the pan or holder of the film should be well insulated. There are reports in the literature which outline the use of relatively high drying air velocity and temperature to ensure that the mass flux or dehydration rate is as close as possible to that of actual spray-drying conditions (Zbicinski and Li 2006).

A useful method currently widely used is to use the single droplet drying approach. The premise of this approach is to dehydrate a single static droplet of the feed material with convective air controlled at a certain humidity, temperature, and velocity. There are many variations in the single droplet drying approach and the main differences between the approaches are in the way the single droplet is suspended. A brief review on the use of the different single droplet drying technique can be found in Sadek et al. (2015). One way to suspend the droplet is to use the acoustic levitation technique. On the mechanical aspect, this form of droplet suspension would represent the most ideal situation, as there is no foreign intrusion into the particles or any surfaces in contact with the droplet.

The acoustic levitation technique utilizes the change in volume to give an indication of the change in mass of the droplet against the drying time (Schiffter and Lee 2007b). Obviously, the diameter change of the droplet can be directly observed. Temperature measurements, due to the noninvasive approach in which there is no physical mechanical intrusion into the droplet, is typically measured by infrared sensors. In using this technique for kinetics measurement, there are certain important aspects to be considered.

Firstly, as the drying rate is inferred from the changes in droplet volume, although the kinetics for the initial wet bulb period can be obtained, the technique might not be able to measure the kinetics of dehydration once the solidification becomes more important. It should be noted that the effect of solute on the drying kinetics normally only becomes significant at the initiation of surface solute concentration increase. Therefore, measurement on the initial period of drying might closely resemble only the evaporation behavior of the solvent. It is unclear at the moment how the falling rate period is fully accounted for. Secondly, in an acoustic levitator, the primary and secondary acoustic streaming will affect the evaporation rate in addition to only airflow convective heat and mass transfer in a typical spray dryer (Schiffter and Lee 2007a). How this additional driving force to the drying process affects the reflection of this technique to the spray-drying process, at the moment, needs further elucidation.

Another single droplet approach is to mechanically suspend the single droplet. This can be done by supporting the droplet on a microbalance (Adhikari et al. 2004). One particular approach is to suspend the single droplet with a glass filament (Chen and Lin 2005). This approach will be described in greater detail mainly due to the familiarity of the author in this experimental approach of single droplet drying. Changes in mass during the dehydration process are measured by observing the minute change in the deflection of the glass filament throughout the drying process, much akin to a fishing rod.

The glass filament approach also allows direct video measurement of the change in droplet size. Temperature measurements are taken by inserting a microthermocouple into the droplet. The glass filament single droplet technique is now commercially available and an illustration of a unit is shown in Figure 5.5. This technique is an intrusive method in which the glass filament is in contact with the droplet. Some analyses have shown that the presence of the glass filament, specifically the mild heat conduction via the filament, does not significantly affect the drying behavior of the droplet.

The main argument often queried by industry pertaining to the single droplet drying techniques, regardless of the way in which the droplet is suspended, is how much does it represent dehydrating droplets in a spray dryer? This is indeed a reasonable question as the size of the droplet used (in the order of millimeters) may be several magnitudes larger than that of

Figure 5.5   Commercially available glass filament single droplet drying rig. (With permission from Nantong Dong Concept Pt Ltd, China— Professor Xiao Dong Chen.)

the sprayed droplets (in the order of hundreds of microns). The dehydration time scales for the single droplet runs are in the order of minutes, whereas the sprayed dried droplets are in the order of seconds. It is important to acknowledge that the single droplet drying approach so far does not provide an exact match to the spray-drying conditions, but mimics and provides an approximation of the drying behavior. At the moment, it is experimentally not feasible to undertake experiments in exact similarity to the scale of the spray-dried droplets. However, drying kinetics derived in these manners, when implemented in CFD simulations, do provide reasonable accuracy in the prediction of the final particle moisture content leaving the drying chamber.

5.4 Effect of Accurately Capturing the Particles Shrinkage and Its Implications

The key physical aspect of particle shrinkage is that the shrinkage behavior is determined by the dehydration history (rate of evaporation) and the amount of initially dissolved solids in the droplet. In the initial period of drying when the dehydration is still in the constant rate period, the shrinkage behavior may approach that of a perfect shrinkage behavior. This is because the droplet is still very liquid-like with no significant shrinkage inhibition effect from the solute or suspended solids in the droplet. For a droplet with initially dissolved solute, as drying progresses, a few possible solidification processes may occur. If the solute in the droplet has a tendency to form a skin or crust layer, the droplet may then experience a rapid decrease in the shrinkage rate even as moisture is progressively removed. The final particle size is then very often determined by the size of the droplet when significant rigidity is developed in the crust. This "crust" size of the particle, analogous to the critical moisture of a material, is also dependent on the drying behavior of the droplet.

For a droplet that does not have the propensity to form a crust, densification of the particle may then occur. In an ideal situation, where the solute solidifies perfectly (fully compacted) such that the density of the solid particle can be described only by the density of the bone dry material, the particle or droplet size throughout the dehydration process can be described by the following expression,

$$d_p = \sqrt[3]{6 \frac{m_p}{\pi} \left(\frac{\rho_w + \rho_s X}{\rho_w \rho_s (X+1)} \right)} \qquad (5.11)$$

This approach has been used in several reports in the literature and also by the author as a simplifying assumption (Woo et al. 2008a,b,c,d, 2011a,b, 2012). Detailed measurements of the shrinkage behavior of some nonskin forming materials have found that the droplet or particle shrinkage may deviate from the perfect shrinkage behavior particularly at high initial solute concentration or towards the lower particle moisture period of dehydration when solidification becomes more significant (Fu et al. 2014). The initial solute concentration also affects the degree of deviation. How is the effect of different initial solute concentrations can be captured in a CFD simulation? One approach is to make use of the linear shrinkage behavior and to further correlate the linear shrinkage behavior to the initial solute concentration.

$$\frac{d_p}{d_{p,initial}} = b + (b-1) \frac{X_{wet\ basis}}{X_{wet\ basis,\ initial}} \qquad (5.12)$$

Where,

$$b = f(X_{wet\ basis,initial}, T_b, \psi_a) \qquad (5.13)$$

An example for milk droplet is shown in Figure 5.6. It should be noted that, although not shown, the shrinkage

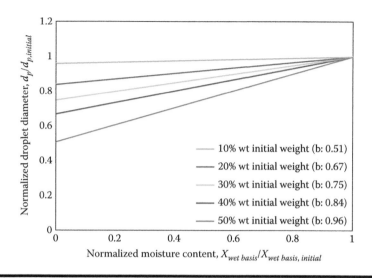

Figure 5.6 Capturing the shrinkage of milk droplet with the linear shrinkage model. (Generated with the equations from Fu, N. et al., *Journal of Food Engineering* 2014, 116, 37–44.)

parameter is a function of the initial drying rate (corresponding to the drying conditions). These parameters are determined for a constant drying condition, in contrast to the changing ambient drying conditions in a CFD simulation. Another peculiar aspect of the linear model is that the final form of the moisture has to be expressed in the %wt for the linear behavior to prevail, expressing the moisture content in the dry basis will result in deviation from the linear behavior.

The diameter of the particle or droplet directly affects two aspects of the CFD simulation. Firstly, the diameter affects the calculation of the drag force exerted onto the particle in the Lagrangian particle-tracking framework. Secondly, the diameter of the particle or droplet affects the computation of the surface area available for dehydration and the calculation of the convective heat and mass transfer coefficient. There is significant nonlinearity between these two parameters as the movement of the particles affects the position of the particle from which the ambient drying conditions affect the dehydration of the droplet. Capturing how the diameter of the droplet or particle changes during dehydration further feedbacks into the computation of the movement of the particle.

When implementing the particle shrinkage model in commercial codes, it is important to implement the shrinkage model in both the Lagrangian particle movement tracking framework and the drying submodel (Yang et al. 2015). This certainly depends on how the particle diameter is numerically tracked in the CFD framework. For example, drawing from the familiarity of the author, FLUENT does not directly track or store the diameter information of the particle. When the particle diameter is required in any subroutine, the particle density is simply calculated using a "perfect shrinkage approach" from the mass and density of the particle which are stored in memory. In such a numerical framework, the computation of the particle diameter following the user-defined shrinkage behavior has to be implemented separately in the Lagrangian particle movement tracking framework and the droplet drying submodel as the particle diameter information may not be directly communicated between the subroutines.

5.5 Capturing the Mass Depression Phenomenon

Most of the reports in the literature utilize the Ranz Marshal equations to describe the heat and mass transfer coefficients. The equations are given in the form below,

$$b_m = \left(\frac{D_{ab}}{d_p} \right)(2 + 0.6Re^{1/2}Sc^{1/3}) \tag{5.14}$$

$$b_m = \left(\frac{k}{d_p} \right)(2 + 0.6\,Re^{1/2}Pr^{1/3}) \tag{5.15}$$

When implementing these models in large-scale CFD simulations, most reports neglect the possible effect of mass flux depression in the droplet drying process. The mass flux depression phenomenon occurs when the high evaporation rate from a droplet (or even flat liquid surfaces) expands the vapor boundary layer adjacent to the surface (Bird et al. 2007). Such an expanded vapor surface will then give more resistance to the transport of evaporated vapor away from the liquid surface, leading to the reduction or depression of mass flow. This effect is often delineated by the reduction in the mass transfer coefficient of the droplet. A survey of the existing correlations to quantify the degree of reduction in the mass transfer coefficient often associates this phenomenon with the evaporation of relative volatile droplets such as fuel droplets typically encountered in the modeling of fuel burning (Renksizbulut et al. 1991; Schwarz and Smolik 1994; Birouk et al. 2008). There are single droplet experimental data and analytical theoretical analyses, however, to show that this phenomenon is also significant for water droplet systems (and flat water film systems), more commonly encountered in spray dryers (Kar and Chen 2004; Chen 2005; Woo et al. 2011a,b). More systematic studies are required to unveil how this phenomenon will affect the overall dryer-wide prediction of the spray-drying process.

5.6 Incorporation of Quality Modeling of the Particles

It is well established that in some material, such as dairy droplets, the initially dissolved solutes will segregate during the convective dehydration process (Faldt et al. 1993). For example, full cream milk powder has excessive overrepresentation of fats relative to the overall bulk composition in the original milk solution. This is because the fat component tends to accumulate onto the surface of the droplet during the dehydration process. Similar observations can also be observed for droplets containing protein and carbohydrate/sugar components. There is abundant evidence to suggest that the material segregation phenomenon is driven by both the surface activity and the dehydration-induced diffusion of the solutes.

Using the distributed diffusion based drying approach as outlined earlier, there have been reports that have investigated and modeled this phenomenon, looking into detail on how the solutes segregate within the "radial shells" of a droplet (Wang et al. 2013). Such an approach needs relatively high computation requirements due to the required discretization of the droplet; however, it provides a very detailed analysis on the segregation phenomenon. On the other hand, based on the diffusion driving force, recent analysis has led to a case-based analytical solution to the material segregation phenomenon (Chen et al. 2011; Jie and Chen 2014; Jie et al. 2015). Both approaches, however, have yet to be incorporated into dryer-wide simulations. If the moisture distribution-based approach were to be implement in a dryer-wide CFD Lagrangian–Eulerian simulation of the spray dryer, in view of the large amount of droplets in a simulation, the high computational requirements have to be overcome. The diffusion driving force analytical models have so far been used to predict the segregated material composition at the "end point" of drying. More work is required if this approach is to be used in a "kinetic" manner.

Chapter 6

Agglomeration and Wall Deposition Modeling

The main motivation to develop and improve the computation of particle–particle or particle–droplet interaction during collision in a dryer wide simulation of the spray-drying process is to predict agglomeration and wall deposition. The prediction of agglomeration within the spray-drying chamber is important because, in the actual process, this affects the bulk density of the spray-dried powder. The bulk density of the spray-dried powder will significantly affect the flowability and the packaging of the product, although this is further manipulated at the second or third stage of drying. Apart from that, agglomerates with different size and mass will traverse across the drying chamber different owing to the different inertia and drag. Another importance of agglomeration within the drying chamber is that it provides an avenue in the elimination of fine particles. These particles are typically separated in the cyclone and returned to the chamber for size enlargement. In view of this, what is actually the current state-of-the-art of agglomeration prediction in spray dryers?

Adopting the EDECAD approach (Verdumen et al. 2004), the physical aspects governing agglomeration can be broken down as follows in a sequential manner: (1) the statistical prediction of particles or droplets colliding with each other during their flight within the drying chamber, (2) the statistical prediction of the collision behavior or dynamics (e.g., head-on collision or collision with tangential motions), and (3) prediction on the outcome of the collision (e.g., whether the particle/droplets stick, coalesce or they rebound from each other).

6.1 Predicting the Collision Efficiency for Agglomeration Application

Modeling on the first two aspects, which are relatively more established, can be found in other fields addressing numerically capturing multiphase particle-airflows (Ho and Sommerfeld 2002; Meyer and Deglon 2011). This short review will, however, focus only on the third aspect in predicting the collision outcome, drawing mainly from the familiarity of the author.

6.2 Modeling Stickiness and Coalescence

6.2.1 Sticking Criterion

For droplets–droplets or droplet–particle collision, coalescence or "fusing" between the droplet and the particles is typically assumed in a dryer-wide simulation. Modeling particle–particle interaction is a little more complex as it is imperative to predict the stickiness of the material. The stickiness of a material, particularly food material, is typically dependent on the moisture content in the particle and the temperature of the particle. These two factors can be combined to delineate the stickiness of the particle with the glass transition temperature of the particle material (Adhikari et al. 2004; Woo et al. 2008a,b,c,d). The glass transition temperature of the particle is highly related to the degree of plasticization within the particle; the higher the moisture content in the particle, the lower the glass transition temperature and vice versa. The glass transition temperature between multiple components (solute and moisture) in a particle can be calculated using the Gordon Taylor equation,

$$T_g = \frac{w_1 T_{g,1} + k_{Tg} w_2 T_{g,2}}{w_1 + k w_2} \tag{6.1}$$

The second component in Equation 6.1 is normally taken as the water component. The k_{Tg} value is specific to the combination of components 1 and 2. If the temperature of the particle is beyond the glass transition temperature (typically higher by 20°C for sugar-based particles), the particle is then deemed sticky. Analogous to this approach, there are dryer-wide simulations that use experimentally measured sticky-curve of the material being spray dried (Harvie et al. 2002). From the author's experience in using the glass transition as the sticking criterion, depending on predicted particle temperature within the chamber, there may be a sudden cutoff in which there is suddenly no particle deposition predicted. It is unclear if such sudden cutoff point in deposition prediction is realistic of not as the sticky point approach does not account for other cohesive/adhesive mechanisms such as electrostatic or van der Waals forces. These other adhesion forces may be significant when the particle temperature is lower than the glass transition temperature or its sticky temperature. In contrast, there is a report in which the van der Waals forces are considered as a sticking criterion for CFD simulation of a pilot-scale counter current tower (Jaskulski et al. 2015).

In contrast to using the sticky point as the criterion, another commonly adopted approach is to use the stick-upon-contact approach; if a particle is simulated to touch the wall, it will be deemed deposited and is removed from the simulation (while releasing its moisture content as mass source into the simulation domain). Some reports in this area are cited here (Huang et al. 2004; Woo et al. 2008a,b,c,d). One may argue about the realism of such an approach as it does not discriminate between a sticky or a nonsticky particle. It has been numerically assessed that usage of the stick-upon-contact approach may lower the average particle moisture prediction at the dryer outlet for an industrial-scale dryer (Jin and Chen 2009). However, an advantage of such an approach is that it provides a standard conservative prediction of the spray-drying operation, without the uncertainty of the more complex deposition model. In the author's opinion, from an industrial perspective, such an approach may be useful as a based case in which very precise computation is not required; for example, for what-if explorations in which only the overall trend on changing various operating conditions is required.

Another approach currently in development, at the time of preparation of this book, is the viscoelasticity approach (Woo et al. 2010). Using the viscoelasticity approach, the collision outcome is instantaneously computed by integrating a piston–dashpot particle contact model, incorporating the storage and the loss modulus properties of the material. This new approach will incorporate the effect of collision dynamics and the rigidity of the particles in determining the collision outcomes. The form of the model for a particle impacting a wall is given below (visualization given in Figure 6.1),

$$F_{collision} = -\frac{A_{contact}}{2r_p}(Ex + \eta v_p) \tag{6.2}$$

Where by geometrical consideration, the contact area can be computed as a function of "penetration" into the wall,

$$A_{contact} = \pi(r_p^2 - (r_p - x)^2) \tag{6.3}$$

The key to this approach is in using the glass transition concept and the William–Landel–Ferry (WLF) viscoelastic shift factor concept to link the particle temperature and moisture content to the modulus properties. Details on this can be

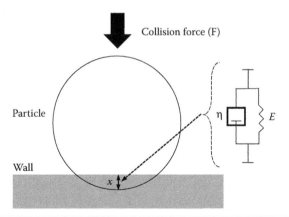

Figure 6.1 **Viscoelastic particle–wall interaction model. (From Woo, M.W. et al. The role of rheology characteristics in amorphous food particle-wall collisions in spray drying.** *Powder Technology* **2010, 198, 251–257.)**

found in the paper cited above. Further work is required to evaluate this new approach in large dryer-wide predictions.

The Ohnesroge number describes the relative significance of the fluid (droplet) viscosity and the inertia and surface tension (Verdumen et al. 2004). This parameter is often used to differentiate between droplet–droplet interactions or viscous (sticky) droplet/particle interaction. Within a CFD simulation, as a droplet moves and undergo dehydration, the viscosity and surface can be computed providing computation of the Ohnesroge number during collision. If the Ohnesroge number is higher than a certain threshold, then it is deemed to be viscosity dominated. Otherwise, it is considered a droplet, which may lead to complete coalescence.

6.2.2 *Important Experiments to Measure the Stickiness of Particles*

It is important to clearly distinguish between the measurement of glass transition and the measurement of sticky point, although they are closely related. This will be particularly important if there is multicomponent segregation in the particle. For example, in milk particles, although the bulk is composed of lactose, the surface is overly represented by fat. While the glass transition property may be more represented by the overall bulk of the particle, the stickiness, which in essence is a surface phenomenon, may be more related to the fats on the surface.

For glass transition measurements, the digital scanning calorimeter is a common analytical technique used. This technique measures changes in the flow of energy through the sample material should any phase change were to occur as the temperature of the material is increased or decreased. Mechanical methods such as the dynamic mechanical thermal analysis (DMTA) are typically used for glass transition measurements of solid or liquid samples. This technique involves rapid oscillating shear to the sample while the temperature of the sample is progressively increased. The idea is to measure changes in the storage and loss modulus. Recent development has allowed such measurements for loose powder using the cantilever DMTA approach. More information can be found in Abiad et al. (2010).

A detailed review on the measurement of stickiness can be found in Adhikari et al. (2001). The viscometry method is a common approach used. The idea is to measure sudden changes in the shear stress of the viscometer when the temperature of the powder sample is progressively increased (Lazar et al. 1956). The tack method involves drying a single droplet and then probing the semi-dried droplet with a force measurement probe (Adhikari et al. 2004). These measurement techniques treat particle stickiness solely as a material property without significant influence on the impacting conditions, although for the tack method, the movement of the probe may affect the measurements.

A unique particle stickiness gun approach was developed by the University of Massey to measure the stickiness of particle as a function of plasticization and impacting conditions analogous to that found in spray-drying chambers (Zuo et al. 2007). In this measurement technique, particles are fed into a convective air stream, which rapidly equilibrates the particles to the airflow temperature, and are then impacted onto a target. The sticky point is then determined when the impact particles start to stick onto the target. In a similar vein, GEA™ has also developed a dynamic tack measurement approach integrating with the single droplet acoustic levitation dryer. The approach allows measurement of particle stickiness during the particle formation process under high-speed collision dynamics.

6.3 Predicting the Structure of the Agglomerate

Once the particle stickiness is determined, the next part in determining the outcome of the collision is to determine the degree of "penetration" between the particles; in other words, how much deformation of the particle occurs in the region of the collision. Some approaches predict the particle penetration assuming the sticky particles to be solely viscous (Verdumen et al. 2004). This approach was implemented and validated only by looking at the final particles size formed, not on the actual penetration between the particles. A more recent approach by the authors considered the possible viscoelastic effect of the sticky particles in such a prediction (Woo et al. 2010). This latter approach, however, has yet to be tested in dryer-wide simulations.

Even if the collision outcome can eventually be predicted, there is a further need to model the development of the structure of the agglomerated particles. At the moment, making reference to the EDECAD project, the "spherical particle" assumption is still used. Specifically, when two particles were numerically determined to agglomerate, the masses of the individual particles were mainly combined and the volume of an equivalent sphere was determined to delineate the "size" of the agglomerate. By this approach, the shape of the agglomerates (along with the distribution of the agglomerate size) which determines the bulk density of the product is essentially "lost" during computation. There is a strong need for more numerical development in this area.

6.4 Capturing the Near Wall Phenomenon for Wall Deposition

6.4.1 *Momentum Impact versus Diffusion Impact*

As discussed in Section 4.3, the Lagrangian–Eulerian approach utilizes the particle-in-a-cell approach where a particle represents a parcel of particles. Although intended as a convenient approach to minimize the computational requirements, as it will be impossible to mathematically capture every single particle in the chamber, it also limits the diffusion movement of particles. In other words, the diffusion of particles is only determined by the random fluctuation of the parcel trajectories and not that of individual particles.

The ability of the Lagrangian–Eulerian approach in capturing the particle diffusion process was evaluated and was found to be reliable in capturing the diffusion process in the "free stream" or the bulk central region of the chamber (Nijdam et al. 2004, 2006). On the other hand, there was limited evaluation of such particle diffusion near the wall. It should be noted that in the real system, particles can reach and collide with the wall via inertial impact (mainly for the larger particles) or via diffusive fluctuation of the particles across the near wall boundary layer. In the absence of such model implementation, one may numerically predict a particle parcel to move very close in parallel to the wall without any particle–wall interaction (Woo et al. 2008a,b,c,d); in the actual process, one might expect some of the moving particles to diffuse due to turbulence and reach the wall. Implementation of such a phenomenon is widely established in more fundamental CFD studies of near wall particles trajectories beyond the boundary of spray dryer simulations. This form of near-wall particle transport modeling was also found in boiler simulations, with the intention of capturing the fouling phenomenon in boilers (Kaer et al. 2006). One report can be found in the literature adopting this approach in a CFD spray dryer simulations (Jin and Chen 2010).

6.4.2 Modeling Particle Removal Due to Shearing

Most CFD work reported in the literature so far does not incorporate the removal of particles. Only one report attempted to model this phenomenon by incorporating deposit removal/reentrainment due to the near wall turbulent burst (Cleaver and Yates 1976; Jin and Chen 2010). The combined approach provided qualitatively realistic deposit flux prediction. Recent work on counter current spraying has shown that the entrainment of particles is significant particular in the spray drying of detergent powder. The removal process is also affected by the impingement of large particles in the air stream. Along that line, continuous drying of the deposits was also found to play a significant role in achieving the final moisture content required for the detergent powder product.

Chapter 7

Simulation Validation Techniques

Ideally, the validation of a CFD simulation of a spray dryer will require comparison between the simulation and the measured (1) velocity flow field, (2) temperature profile of the air, and (3) humidity profile of the air within the chamber. In addition, comparison should also be made at the outlet from the dryer by measuring the (1) product yield, (2) product moisture content and size distribution, and (3) outlet air humidity and temperature. Lastly, the deposition fluxes predicted by the simulation should also be verified experimentally.

Obtaining a complete set of this information from an industrial spray dryer is difficult, particularly in measuring the internal airflow profiles and the temperature and humidity distribution. Most industrial validation is limited to comparing the prediction of the simulation at the outlet (airflow measurements and product measurements). It is typically assumed that if the predicted outlet conditions are in agreement with those observed from the actual operation, the prediction of the internal of the spray dryer "should" be correctly predicted. This is a simplistic but certainly a practical assumption. Even if a simulation is fully validated, in the opinion of the author, there is still a possibility that multiple errors in the simulation "balance" each other out, producing an apparently "good" prediction. Therefore, the modeler should always evaluate if the airflow pattern predicted within the dryer is physically logical or not. This can be undertaken by drawing comparisons with measured observations, which are possible in smaller laboratory or pilot scale spray dryers.

How much validation is required for a CFD model of a spray dryer? One of the main applications of a CFD simulation is to explore what-if situations. It defeats the purpose to undertake full validation for all the simulation cases to be explored. Nevertheless, before utilizing any developed CFD model of the spray dryer, it will be useful to validate firstly the "dry-run" case without feed spray. This will provide the first level of validation without the complexity or possible errors associated with the introduction of sprayed droplets in the simulation as well as in the actual operation. Once the airflow only simulation is verified, a few base cases with feed spray have to be validated. Ideally, on top of showing that the simulation results provide a reasonable match to the operation data, these runs should be planned such that they can also be used to assess how well the model response to changes in the operating conditions. For example, when the outlet air temperature is controlled at a higher level, the inlet air temperature will be increased. Is the model responding reasonably to the same trend and magnitude? Some of the important parameters to be varied are the change in feed spray rate and the control of the outlet temperature.

7.1 Airflow Measurements

Quantitative measurements reported in the literature so far are limited to pilot-scale and laboratory-scale spray dryers. Hotwire measurement is a common technique reported by many papers in the literature (Kieviet and Kerkhof 1997;

Woo et al. 2009a,b). The hotwire measurement involves placing a heated element in the flow field. The cooling effect on the wire induced by the flowing air changes the resistance in the wire. By correlating the changes of the resistances to the amount of cooling in the wire, the velocity of the flow field can then be computed. Commercially available hotwire measurement anemometer is precalibrated for such measurements. This technique is now commonly used in industrial airflow measurements in ventilation ducts or large airflow lines. More elaborate and more expensive high spatial resolution units are also used in detailed boundary layer flow study.

For spray drying airflow applications, a few important factors have to be considered when considering hotwire measurements:

1. Hotwire measurements can only be undertaken in "dry run" conditions in which there are no sprays. This is because collision of particles onto the delicate wire of the hotwire anemometer may damage the wire or its deposits onto the wire may affect the interpretation of the measurements. In view of this, if used for validation purpose, the measurements may only be useful for the first stage of validation for the airflow only.
2. Although the hot wire is normally designed to be heated to significantly high temperatures to facilitate any possible cooling for the velocity measurement, the holder of the housing may not be able to withstand the high temperature typically found in spray dryers. From the author's experience this is particularly true if industrial hotwire units are used. Therefore, hotwire measurements may be more suited for cold dry runs without heating to the inlet air or if mild air temperature is used.
3. Hotwire anemometers with a single wire are designed to measure the magnitude of the airflow perpendicular to the wire (or the equivalent magnitude of the airflow). Therefore, the direction of the velocity measured is

dependent on the position and orientation of the hotwire element. Industrial hotwire units are typically fitted with a sheath or protective cover to minimize nonperpendicular airflow across the wire. This is an important consideration especially if significant swirl is present in the flow field. More complex laboratory hotwire systems may utilize multiple hotwire elements, which allow measurement of velocity components providing three-dimensional information of the flow field.

4. The sensitivity and the frequency of the hotwire measurements will determine if turbulence properties can be measured or not. Low measurement frequency in industrially available units typically does not allow measurement of the turbulence fluctuations in the airflow, which limits the data available for full comparison. High-frequency measurements are available in more expensive laboratory hotwire units. Selection between the two will certainly depend on the purpose of the measurements. In addition to validation purposes, turbulence measurements will be useful for understanding the dispersion of particles in the system.

5. Access and positioning of the hotwire probe is perhaps the biggest limitation in measuring the internal airflow behavior of a spray dryer and perhaps explains why measurements reported hitherto is limited to pilot-scale and laboratory-scale spray dryers. Most of the reports, including that by the author, resorted to modification of the viewing port for positioning of the hotwire probe into the spray dryer (Kieviet and Kerkhof 1997; Woo et al. 2009a,b). Such a modification may not be suitable for large commercial-scale spray dryers.

In place of hotwire measurements, pilot tubes may also be used for internal airflow measurement. Although not reported in the literature, from the author's experience in using the pilot tube, the measured velocity is significantly sensitive to the orientation of the pilot. At the same position within the spray dryer, slight discrepancies in the orientation of the pilot tube will result in significantly contrasting velocity measurement. From a practical viewpoint, the hotwire measurement is less susceptible to such slight discrepancies; difficulty in access may prohibit complete elimination of such orientation discrepancies. Particle image velocimetry (PIV) measurement is also another method commonly used to quantify airflows for validation with CFD simulations.

Qualitative observation on the airflow pattern in the spray dryer can be undertaken by visual observations of cotton turfs installed into the spray dryer (Southwell and Langrish 2003; Woo et al. 2009a,b; Gabites et al. 2010). Other materials such as light fabric pieces can be used; however, for discussion in this section, the term cotton turfs will be used. The idea is that the direction of the movement of flapping of the cotton turfs will give an indication of the flow direction and fluctuation. From the author's experience, very light material has to be used, especially if the technique is to be utilized to delineate the flow pattern in regions within the drying chamber away from the central core jet airflow. Such visual observations are limited to "dry runs"; however, in contrast to the hotwire measurements, hot airflow conditions can be evaluated, provided the cotton turfs can withstand the high-temperature conditions. This technique was also successfully used to understand the flow pattern of an industrial spray dryer with a bottom static fluidized bed (Gabites et al. 2010).

Apart from that, the smoke technique can also be used to visualize the flow pattern within the drying chamber (Langrish et al. 1992; Southwell and Langrish 2003). This technique involves introducing smoke into specific regions of the drying chamber and observing flow delineated by the movement of the smoke. It should be noted that the smoke is not introduced at the inlet air stream into the spray chamber. The smoke visualization technique is commonly used in fluid mechanics and wind tunnel study and was successfully used in a report on spray drying flow observations (Southwell and Langrish 2003). Based on the author's experience in using this technique for a pilot-scale spray dryer (unpublished data), one potential challenge in using this technique is the possible rapid dispersion of the smoke upon introduction into the chamber, which might hamper flow visualization. The spray dryer, which was available to the author, was one fitted with a rotating atomizer and had significant swirl (Woo et al. 2007a,b). Nevertheless, such experimental challenge may be dryer specific depending on the degree of turbulence in the drying chamber.

7.2 Temperature and Humidity Measurements

Thermocouples and humidity probes can certainly be positioned within the spray chamber for temperature and humidity measurements of the airflow. Conventional temperature and humidity probes can only be used in "dry run" conditions because any particles or droplets coming into contact with the probe will affect the measurements. The challenge is in making such measurements in the presence of sprayed droplets or particles during the actual spray drying operation. Why is it important to make these measurements? The difference in the temperature, between the dry runs and the operating with sprays, will delineate the experimentally measured rapid evaporation region within the spray chamber. In this way, in addition to providing very valuable information for validation with the CFD simulations, knowledge on the rapid evaporation region can also be used to evaluate the effectiveness of the spray drying process designed in utilizing the chamber space for evaporation.

The group at Eindhoven University of Technology has developed a unique airflow sampling probe which allows separation of the droplets or particle (~10 μm lower threshold) from the airflow prior to reaching the temperature- and humidity-sensing elements (Kieviet and Kerkhof 1997). The working principle of the design is in aspirating the particle-laden airflow from the spray dryer (the probe being installed to specific locations within the chamber) and then abruptly reversing the airflow into a smaller region located within the larger probe. In this way, particles will not be able to follow the sudden airflow curvature and become separated from the airflow. The "clean" airflow is then passed on to temperature and humidity sensors for measurement. This technique was reported for use in a pilot-scale dryer, and to date the experimental data obtained via this technique remain very highly cited. The simplicity of this design will allow relatively easy setup of the measurement probe.

7.3 Yield, Product, and Deposition Flux Measurements

Product yield is simply determined by measuring the amount of powdered products obtained relative to the amount of dissolved solids sprayed. For the laboratory- and pilot-scale spray dryer, this can be collected from the outlet cyclone of the dryer. For industrial-scale dryers, this can be collected from the outlet of the external fluidized bed (should it be a two- or three-stage drying process). The potential difficulty for the industrial-scale dryer is that fine particles normally leave the chamber via the airflow, whereas the larger particles exit the dryer via the bottom exit. Although all the fines are normally returned into the chamber or eventually returned into the final product at the fluidized bed, for validation purpose, it may be useful to discriminate what proportion of the final product is from the fine exit and from the bottom larger particle exit, respectively. This aspect has to be kept in mind when undertaking simulations for the industry, as currently the fluidized bed is typically not part of the spray drying simulation; only the fluidized bed airflow is accounted for at the time of writing.

Similarly, for an industrial system, the same challenge will be present in determining the moisture and the particle size distribution of the product. For small laboratory-scale or pilot-scale spray dryers, the moisture content can be determined from the thermogravimetric method. Particle size analysis can be undertaken from laser diffraction measurements. For industrial systems, accurate measurements for these two parameters may be difficult as the secondary- or third-stage fluidized bed operation may further reduce the moisture content of the particles and may further induce agglomeration of the particles, affecting the particle size measurement. One possible approach is to directly collect the particles by "by-passing" the fluidized. However, in certain system, effective operation of the cyclone in removing the fines (which returns the particles to the fluidized bed) may rely on maintaining the flow pressure in connection with the fluidized bed unit. Solutions to moisture content measurements for an industrial system are case specific and these have to be clearly communicated to the industry so that expectations in the validation of the simulations can be managed.

Most deposition flux measurements have so far mainly been reported for pilot-scale and industrial-scale dryers (Ozmen and Langrish 2003b; Kota and Langrish 2006; Woo et al. 2008a,b,c,d, 2009a,b). This was done by hanging or fixing plates at different positions on the spray dryer chamber wall. Figure 7.1 illustrates one method of inserting such plates in a pilot-scale dryer. There was also one group at the University of Leeds focusing on collecting and analyzing deposit fluxes in a counter current spray dryer producing detergent powders (Francia et al. 2015). In one specific report, for an industrial-scale dryer, the deposits were collected with a unique rolling apparatus and the deposition behavior, specifically the wavy deposit pattern at the conical–cylindrical wall interface of the drying chamber was observed (Chen et al. 1993). Such a wavy deposition pattern was also observed in pilot-scale dryers (Woo et al. 2007a,b) delineating the generality of the

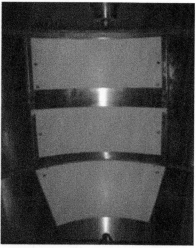

Figure 7.1   Insertion of collection plates into a pilot-scale spray dryer for deposition flux measurement. (From Woo, M.W. et al., *Drying Technology* 2008b, *26*, 1180–1198; Woo, M.W. et al., *Drying Technology* 2007a, *26*, 15–26.)

observations and possibly denotes the transient behavior of the air particle flow behavior in a spray dryer. Such a wavy deposit pattern is illustrated in Figure 7.2, and may also be used to qualitatively compare with the simulations.

Deposition flux measurements will be useful to partly validate the tracking or trajectory prediction of the particles in the CFD simulation. When making such measurements, a few important aspects should be noted:

1. The material of the collection plate (particularly differences in surface roughness and surface energy) will significantly influence the deposition flux, particularly if the deposit involves semi-wet particles.
2. The rate of particles depositing initially on a clean wall (adhesion) will be different from subsequent particles deposition on already adhered particles on the wall (cohesion). For a better representation of a steady-state operation of a spray dryer, it will be important to analyze

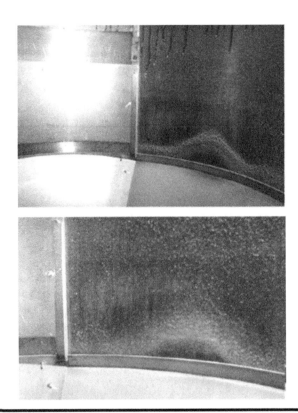

Figure 7.2   Coherent wavy deposition behavior in a pilot scale spray dryer. (From Woo, M.W. et al., *Drying Technology* 2007b, *25*, 1741–1747.)

measurements only when the deposition behavior is well into the latter period; unless the measurement is to investigate the initial deposition behavior in detail. At the time of writing, deposition-modeling submodels have yet to clearly discriminate the adhesion and cohesion phenomenon.

3. From the author's experience (Woo et al. 2008a,b,c,d) and also that reported by Kota and Langrish (2006), deposition flux measurement at the conical region of the spray chamber may be affected by random avalanche removal of particles. This region of the spray dryer chamber wall involves deposition of relatively dry particles, which are

weakly attached to the wall. Furthermore, the deposition flux would be relatively high due to the typical high air curvature at that region.

7.4 Controlled Experimental Technique for Model Development

There are relatively few reports in the literature with detailed measurements of the internal airflow or observation of the airflow pattern in the chamber. Table 7.1 summarizes some of

Table 7.1   Detail In-Chamber Experimental Measurements and Observations Reported

Authors	Spray Dryer Details	Measurement Details
Kieviet and Kerkhof (1997)	Pilot-scale spray dryer with nozzle atomizer	Humidity, temperature, and velocity measurements
Gabites et al. (2010)	Industrial-scale two-stage unit	Tell-tail pole airflow visualization
Southwell and Langrish (2003)	Pilot-scale spray dryer	Cotton tuft and smoke visualization
Bayly et al. (2004)	Counter current spray dryer	PIV velocity measurement
Wawrzyniak et al. (2012)	Counter current spray tower	PIV velocity measurement
Fieg et al. (1994)	Industrial spray dryer	Velocity and temperature profile
Zbicinski et al. (2002) Zbicinski and Piatkowski (2009)	Large-scale counter current spray tower	Laser Doppler anemometry (LDA) and particle dynamics analysis (PDA), humidity, and temperature measurements
Woo et al. (2009a,b)	Pilot scale unit with rotating atomizer	Hotwire measurement and cotton turf visualization

the reports that are available in the literature. It is noteworthy that the modeler will still need to validate their specific spray drying simulation. These data, however, will be useful for validation of a base case to provide a basis for numerical developmental work such as the development of a particular submodel or a new numerical scheme etc.

In a spray dryer, particles with different sizes will tend to follow different paths or trajectory in the drying chamber. This is because of the different inertia of the particle and the different drag force exerted onto the particle by the air. In fact, particle segregation within the spray chamber has been observed experimentally in spray dryers (Woo et al. 2008a,b,c,d). Such a phenomenon will lead to different drying history experienced by the particles. This then adds a layer of complexity in validating any droplet drying or particle quality predictive model

Figure 7.3   Commercially available monodispersed spray tower. (With permission from Nantong Dong Concept Pt Ltd, China—Professor Xiao Dong Chen.)

developed for the CFD framework. Adding more ambiguity is that product moisture or quality can only be measured from the bulk powder collected at the outlet, which does not discriminate any difference in particle sizes within the sample.

One option to remove this layer of complexity is to use monodispersed (same sized) droplets from the atomizer. By using monodispersed droplets, each droplet will consistently experience the same drying history within the chamber. This control can be further enhanced by allowing the droplets to travel in a one-pass manner in a tall-form drying chamber, by a combination of low air velocity and allowing the droplet to fall mainly driven by gravity pull, minimizing possible recirculation of the particles. Figure 7.3 shows such a monodispersed spray tower, which can be commercially available. The spray tower utilizes one or multiple monodispersed nozzles, which can produce droplets in the range of the 40–200 μm. Monodispersed droplets are generated from a proprietary piezoelectric nozzle (Wu et al. 2011). On top of using this experimental technique for model development validation, in view of the well-controlled consistent drying history of the droplets, is also used for product and formulation development (Liu et al. 2012; Wu et al. 2013).

Chapter 8

Common Challenges for Industrial Applications

8.1 Managing Expectations

If the CFD technique is introduced into the management of a spray-drying facility, it is very common that the technique is expected to solve "all" the problems encountered in the spray-drying process. It is important to note that the accuracy of the simulations is highly dependent on the information that can be supplied to the modeler. Although such information can be easily measured and obtained in the small or pilot-scale laboratory environment, the challenge in using this technique for industry is in obtaining such information. One main challenge arises because the modeler and the industry partner often view the process in different lights. To be more specific, the information required by the modeler is very often different from what is required to operate a spray dryer. This will be discussed in detail in the following sections, with the aim of bridging this gap. In view of this challenge in communicating with industry, it was found to be useful to provide a template to industry to indicate the requirements for a spray dryer simulation.

In some cases, getting this information might not even be possible, and this will jeopardize the accuracy of the simulation. However, the management always expects the simulations to be used with high confidence. Therefore, in such situations, it is pertinent that the modeler manages the expectations from the simulations and also adopts some strategies that will allow interpretation of the results incorporating the uncertainty involved. Below are a few common scenarios typically encountered when interfacing with industry, collated from the author's experience.

8.1.1 Estimation of Droplet Size for Atomization of Concentrated Feed

This is by far the most commonly encountered challenge when using the CFD technique for the industry. It should be noted that the size and distribution of the initial droplets incorporated into any CFD simulation are the key criteria in the prediction of the process. A sensitivity study on a well-controlled mono-dispersed spray-drying system (unpublished results by the author) revealed that even a variation of 20 micron in droplet size is sufficient to significantly alter the prediction of the drying behavior in a spray chamber. This is particularly significant if the droplet size is large and may be kinetically limited in drying, and if there is insufficient time to reach equilibrium moisture before exiting the dryer or hitting the wall or perhaps colliding with other particles or droplets. Therefore, it is very important to get this input into the simulation accurately.

One common challenge is that while the modeler requires droplet size distribution, the industrial operator although aware of the importance of the droplet size distribution to the process, typically works with other parameters such as the atomizer pressure, rotating speed, feed rate, orifice diameter, etc. This is fully understandable as these are the parameters that can be directly varied in a spray dryer simulation to affect atomization; however, they are not directly useful or required in a simulation.

One suggestion to overcome this is to go back to the information provided by the vendor of the atomizer. Common nozzle or rotating atomizers used in the industry are normally calibrated with water. If the feed material has the density, viscosity, and surface tension property similar to water, then the calibrated droplet size distributions provided by the atomizer supplier can be used. Such an approximation is typically applicable for very dilute spray feed solutions.

However, in most industrial applications, the feed solutions are far from similar to the consistency of water. Commercial economics have driven spray dryers to handle high concentrates; in fact, the higher the concentrate, the better. In spray drying of dairy production, feed concentration can reach as high as >50% wt dissolved solids. Furthermore, some of these materials may contain additional suspended nonsoluble fats, making them deviate further from water. Some spray-drying process, for example, in the production of catalysts, may also consist of suspended solids, which alters the flow behavior of the fluid upon atomization, deviating from the behavior of water. In such situations, usage of precalibrated atomized droplet sizes may not offer sufficient accuracy. To compound this challenge further, CFD simulations are typically set up to evaluate different spray rates and atomization pressure or rotations. Different spray rates or atomization potential will alter the size distribution of the droplets.

Measuring the atomized droplet size distribution may be difficult for industry. Even if a laboratory setup can be used to measure the atomization parameters by laser measurements or other technique, sanitary or production requirements may not allow the removal of the atomizer lance (the nozzle and the accessories) from the production facility. Furthermore, such a measurement facility may require setting up of a feeding system to delivery high-pressure pumping in the case of measurement on a nozzle atomizer.

One suggestion is to resort to empirical correlations available in the literature. Some commonly used empirical correlations available are given by Masters (1979). These correlations will be able to account for changes in the feed rate, atomization potentials and geometry or the atomizer. Some of the correlations are also available and are used in commercial codes such as FLUENT. When using these empirical correlations, it is important to note that the empirical nature of the correlations means that they are specific to the atomizer from which they were developed. The author has experience in using rough back-calculation of the droplet sizes from the product particle size distributions, showing that the correlations may predict an erroneous droplet size distribution (Woo et al. 2008a,b,c,d). Therefore, these correlations may be used as a guide and comparison has to be made with the final particle size distribution produced as an indication on the accuracy of the applicability of the correlation specific to the atomizer used.

If there is such an inherent uncertainty in the atomized droplet size distributions of the atomizer used, can we use the CFD simulations? If accurate information on the size distribution of the atomized droplet is not available, one suggestion is to use the correlations available or to estimate a droplet size distribution from the product as a basis and then undertake simulations incorporating the upper bound and lower bound of the simulations. The upper and lower bound can be determined by adjusting the mean droplet diameter, the spread of the distribution, and the largest and lowest droplet size in the distribution. In such an instance, the CFD simulation will then be used to predict a trend on how changes in a combination of parameters will change the behavior of a particular spray-drying process. Such an analysis can then be used to provide an operation window for the spray dryer of interest. Direct interpretation of the "exact value" from the simulations should then be avoided and this should be made clear to the management so that the expectation can be effectively managed.

8.1.2 Complex and Lack of Information on Air Inlet Configurations

Capturing the geometrical information of the air inlet region is very important as it affects prediction of the initial droplet–air contact. This will significantly affect prediction on the initial phase of drying where significant quality and physical changes are expected to occur. When communicating with a plant operator or engineer, at the mention of geometry information of the region of airflow inlet, very often, information will be given on the plenum chamber external to the drying chamber. Such information is typically not essential in a typical simulation of the spray chamber, as the simulation domain only constitutes the chamber in which the droplet is atomized. This was illustrated in Figure 4.4. This is unless detailed simulation is required on the air distributor, which is normally undertaken as a separate simulation. However, one has to evaluate the need for such fine details relative to the computational requirements and the aims of the simulation. In most cases, simplifications can be made. One key simplification to capture is whether the air coming into the chamber is straightened or swirled with an angle.

Apart from that, air may not enter the simulation domain from only one source. It is important to accurately capture all the air sources that may affect the initial droplet–air contact, regardless of their airflow magnitude. In spray dryers with nozzle atomization, the nozzle is often positions into the spray dryer with an atomizer lance. The atomizer lance may consist of an annulus cylinder, which may constitute the cooling air and, although detailed information of such a lance is good to know, what is important is how the lance is obstructing the airflow in the chamber and the location in which the air enters the simulation domain; the regions within the lance are typically not required in the simulation.

8.1.3 Is the Spray Dryer Well Insulated?

Most industrial spray dryers may seem to be well insulated. However, there is bound to be heat loss from the chamber. Such heat loss may not significantly affect the drying behavior in the region of the initial droplet–air contact; however, it may significantly affect the temperature of the drying air at the outlet of the chamber. For large spray dryer (or for operations with smaller particles) in which the particles have sufficient time to reach the equilibrium moisture conditions, the outlet conditions play a significant role in the prediction of the final particle moisture leaving the spray chamber. Therefore, it is very important to account for such heat loss. One approach, following Langrish and Zbicinski (1994), is to approximate an overall convective heat loss coefficient for the chamber. This approach, of course, only applies if heat loss is modeled to be lost from the chamber by convection. An advantage of this approach, in contrast to specifying a value of percentage of heat loss, is that heat loss can then be calculated accounting for different external ambient conditions; that is, spray drying in winter or summer, etc. How do we then approximate an overall heat loss coefficient?

One suggestion is to simulate a "dry run" in which only hot air is introduced into the chamber and allowed to flow and exit the chamber. The inlet and outlet air temperature will then need to be compared to similar "dry run" operating conditions in the actual spray dryer. A suitable overall heat transfer coefficient which allows the simulated temperature drop to match with that observed from operation, can then be evaluated. An average heat transfer coefficient to account for the loss can be used for the entire chamber. This approach has been evaluated for several pilot-scale spray dryers and provided reasonable agreement (Woo et al. 2008a,b,c,d, 2011a,b).

Where do we then get such "dry run" operation data? This may be obtained during the startup heating up period of the spray dryer before the initiation of any spray. It may be tempting to use the inlet and outlet temperatures observed during the initial water spray which is typically used before switching over to the more concentrated feed spray; this is a convenient measure that can be supplied by industry. Considering that the dry run was intended to produce a condition within minimal ambiguity, one has to think about the accuracy of such an approach. A common argument is that as pure water evaporation modeling is relatively well established, using such a condition as the controlled case is suitable for obtaining the heat loss coefficient. It should be noted, however, that the introduction of sprays regardless of water or actual concentrated feed already introduces a degree of uncertainty into the simulation because of possible assumptions in the atomization parameters. Therefore, the resultant fitted heat loss parameters may already be augmented by such uncertainty and may or may not be a good representation of the actual heat loss in the system.

8.1.4 How Do CFD Simulations Tie in With Plant-Wide Prediction Packages?

Industrial operators may often expect the CFD simulations to be a part of the overall process simulation tool. Some of the plant-wide analysis tools available are Aspen, Symprosis, etc. Such expectations may stem from the need to operate and optimize a whole plant operation. Furthermore, to a great extent, the capacity of a spray dryer is tied in with the evaporators upstream. In terms of the spray-drying operation line, an industrial operator may be interested in optimizing the pump capacity and the return line or energy requirements, etc. It is important to clearly specify that the CFD technique is mainly used to examine the process within the spray-drying chamber and is not a primary tool as part of the processing line analysis. There were indeed suggestions to incorporate the predictive outcome from a CFD simulation into a spreadsheet-like or plant-wide simulation platform. This is indeed possible technologically as it will involve mainly communication between the CFD simulation and the plant-wide simulation platform. However, the simulation time scale between the two may be significantly different. A plant-wide simulation may take a few second or at most minutes to compute, whereas a CFD simulation may take hours or even days to compute depending on the complexity of the spray-drying process. In the opinion of the author, a plant-wide heat and mass balance simulation might not require the detailed or accuracy provided by a CFD simulation and more simplified spreadsheet-like one-dimensional (Patel et al. 2010; You et al. 2014) computation or equilibrium-based computation (Ozmen and Langrish 2003a,b) or effective rate approach (George et al. 2015) that takes seconds to compute may already be sufficient.

This raises another question: Can CFD be used as a routine plant-wide operation training tool? Similarly, the focus of the CFD technique is to allow understanding and visualization of the internal of the spray-drying chamber. Even if the CFD tool is to be used solely for training on the operation of the spray dryer, the relatively long computation may not lend itself to be used as a "live" tool to illustrate what-if situations. A set of preset calculated results may need to be undertaken *a priori* to any training. In this respect, the one-dimensional simulation approach may be more suitable as a training "live" training tool, provided the airflow of the spray dryer lends itself to a one-dimensional plug flow analysis. At the moment, spreadsheets have been developed for long tower spray dryers, counter-current and cocurrent, but there is yet to be a one-dimensional predictive approach for airflow pattern that is recirculated within the spray-drying chamber.

8.1.5 I Want to Understand Why My New Spray-Dried Formulation Is Off-Specification!

The statement above is certainly a rhetorical one, but represents some of the expectations that come across from industry (or even academic researchers) when introduced to the CFD technique for spray dryers. This may be in preventing protein denaturation in particles, maintaining minimal particle color changes induced by spray drying or perhaps to maintain probiotic survival during spray drying, etc. It is true that the CFD technique is introduced as a tool to solve industrial problems or to improve on a process. What is important to know is that for industrial applications, the CFD technique can be used mainly to provide a better understanding on the effects of operational and chamber geometrical conditions on the spray-drying process. It cannot be used to provide a fundamental understanding of how external drying conditions affect the quality changes in the particles or droplets during dehydration. Those fundamental understandings have to be well established mathematically (submodels as illustrated in Chapter 1) and are actually input parameters into the CFD simulation, should such quality changes be investigated. The CFD simulation is then used only to predict how the quality changes manifests, based on the well-established models, corresponding to the predicted drying conditions within the chamber.

Take the following case in point. A company wants to use CFD simulations to understand how protein denaturation occurs in their spray-dried formulation under different spray-drying conditions. The first step to undertaking such a CFD simulation endeavor is actually not the simulation but to firstly understand and obtain the denaturation kinetics or model of the protein corresponding to different ambient heating conditions. This is because the basic CFD simulations can only provide information of the drying conditions experienced by the droplets or particles. How these local drying conditions affect the particle denaturation process certainly depends on the denaturation kinetics incorporated into the basic CFD frame-work. Requirements for such kinetics are given in Section 5.5. This is an important aspect of a CFD simulation project as it will determine how an entire CFD project is scheduled or costed. The "non-CFD" part of the work in determining the quality prediction submodels may actually require significantly more time and resources when compared to the actual CFD simulation components.

8.1.6 How Does the CFD Model Reflect My Actual Feed Material?

This is a very common question asked by the industry. Although such details are easily comprehensible by the modeler, it is very important for the modeler to educate the industry partner or management on how this is captured mathematically. In essence, the different feed materials are captured by their different physical properties and the drying behavior. In the simulation, the physical property of the feed material will be characterized by the density, initial solute concentration, the specific heat capacity and the latent heat of evaporation of the solvent involved. In the context of spray solidification modeling, the latent heat of fusion then becomes another important parameter. In addition to the physical properties, different drying kinetics corresponding to the feed material will also be incorporated.

Another question often beckons: How about the proportion of fat, protein, or carbohydrate, etc. in the feed material? How are they accounted for? These specific details need not be incorporated directly into the simulations as they are mainly used to compute the physical properties of the feed materials and the drying behavior. Therefore in the "eyes" of the computer, the simulation does not recognize what type of feed material is used but only recognizes the difference in physical properties and different drying behavior.

The next question, which typically ensues, is as follows: Will the simulation distinguish the different quality properties of the feed material? At the time of writing, the quality parameters are typically not used to distinguish the different materials. This is because the current development of the CFD simulation technique for spray dryers employs a one-way coupling between the drying behavior and the development of quality parameters; the drying behavior affects the development of the quality parameters and no feedback from the quality parameters into the drying behavior. For example: The CFD simulation, based on the drying kinetics implemented, predicts a very rapid dehydration from the semi-dried particles. From the dehydration rate, the protein denaturation model (if implemented) predicts a certain degree of protein denaturation in the particle. The degree of denaturation predicted, however, is typically not fed back to the drying model to augment the drying behavior; even if the denaturation may have the possibility of augmenting the drying behavior.

Such one-way coupling between the drying kinetics and the quality parameter prediction is strictly limited by the developments of these submodels. Nevertheless, at the current state of development, the drying behavior and the physical properties of the feed material are still the two parameters used to distinguish different feed materials.

8.1.7 Strategy for Outlet-Controlled Spray Dryers?

This last section of the book was written more for the modeler than for the industrial counterpart. Nevertheless, for the industrial counterpart, this may serve to illustrate the limitation or difficulty involved in CFD modeling of certain configurations of industrial-scale spray dryers. Some industrial-scale spray dryers are mainly controlled by specifying the required outlet temperature of the chamber and in some cases, also partly controlled by specifying the internal pressure of the chamber (mainly slightly vacuum). When the feed spray rate and atomizer used in the process are fixed, the system will then be automatically controlled by adjustments to the outlet blower and the inlet heating temperature of the air. The adjustment to the outlet blower is mainly to control the pressure within the chamber, whereas the temperature of the inlet air is used to maintain the outlet temperature from the chamber. In such a system, the mass flow rate of the air may not be measured.

Most reports in the open literature in the CFD simulation of spray dryers are undertaken with input information in the air inlet conditions with the broad aim of predicting the outlet conditions and subsequently the particles' conditions at the outlet. For the type of industrial system discussed above, the requirement from an industrial counterpart part may be different. It could be that given a set of desired outlet conditions, the emphasis will be on what suitable inlet conditions are to be used to maintain the outlet conditions, accounting for the feed spray rate and the atomization characteristics chosen. Such a simulation may be numerically challenging and may not be directly implementable in a commercial code. This is because in a CFD simulation, it is numerically "more natural" to predict the flow field following the direction of the flow instead of working backwards. Incorporation of such a backwards computation may also be more challenging when coupled with droplet drying in the Eulerian–Lagrangian framework.

8.1.8 How Accurate Is the Final Particle Moisture Content Prediction?

Requirements on the accuracy of the final particle moisture prediction certainly depend on the intended application of the model. In the author's opinion, if the CFD technique is applied for industrial driers, the accuracy may be in the range (or order) of ±1% or 2% wt moisture content (or maybe even larger range) if correctly set up. Such an accuracy level may be suitable for troubleshooting what-if scenarios or in dryer design to predict the change in dryer performance when certain parameters are changed. In such a situation, the CFD simulation can be used to give an indication on a possible range of suitable operating conditions; this will be particularly useful when spray drying new formulations without *a priori* experience. However, the current level of accuracy with the CFD technique for spray dryers may not lend itself to very detailed fine-tuning or control of existing spray-drying operations. It should be noted that the source of uncertainly, as described earlier, lies in obtaining reliable boundary conditions of information for the model and is not due to major discrepancy or limitations of the models employed.

References

Abiad, M.G., Campanella, O.H., Carvajal, M.T. Assessment of thermal transitions by dynamic mechanical analysis (DMA) using a novel disposable powder holder. *Pharmaceutics* **2010**, *2*, 78–90.

Adhikari, B., Howes, T., Bhandari, B.R., Truong, V. Stickiness in foods: A review of mechanisms and test methods. *International Journal of Food Properties* **2001**, *4*, 1–33.

Adhikari, B., Howes, T., Bhandari, B.R., Truong, V. Effect of addition of maltodextrin on drying kinetics and stickiness of sugar and acid-rich foods during convective drying: Experiments and modelling. *Journal of Food Engineering* **2004**, *62*, 53–68.

Bayly, A.E., Jukes, P., Groombridge, M., McNally, C. Airflow patterns in a counter-current spray drying tower—Simulation and measurement. *Proceedings from the International Drying Symposium* **2004**, pp. 775–781.

Bird, R.B., Stewart, W.E., Lightfoot, E.N. *Transport Phenomena*. 2nd edition. **2007**. New York: John Wiley and Sons Inc., Chapter 22, pp. 703–710.

Birouk, M., Maher, M.A.A., Iskender, G. Droplet evaporation in a turbulent environment at elevated pressure and temperature conditions. *Combustion Science and Technology* **2008**, *180*, 1987–2014.

Chen, X.D. Lower bound estimates of the mass transfer coefficient from an evaporating liquid droplet—The effect of high interfacial vapor velocity. *Drying Technology* **2005**, *23*, 59–69.

Chen, X.D. The basics of a reaction engineering approach to modelling air-drying of small droplets or thin layer materials. *Drying Technology* **2008**, *26*, 627–639.

Chen, X.D., Lake, R., Jebson, S. Study of milk powder deposition on a large industrial dryer. *Transaction of IChemE* **1993**, *71(c)*, 180–186.

Chen, X.D., Lin, S.X.Q. Air drying of milk droplet under constant and time-dependent conditions. *AIChE Journal* **2005**, *51*, 1790–1799.

Chen, X.D., Putranto, A. *Modeling Drying Processes—A Reaction Engineering Approach*. Cambridge University Press, UK, **2013**.

Chen, X.D., Sidhu, H., Nelson, M. Theoretical probing of the phenomenon of the formation of the outermost surface layer of a multi-component particle, and the surface chemical composition after the rapid removal of water in spray drying. *Chemical Engineering Science* **2011**, *66*, 6375–6384.

Chew, J.H., Fu, N., Woo, M.W., Patel, K., Selomulya, C., Chen, X.D. Capturing the effect of initial concentrations on drying kinetics of high solids milk using reaction engineering approach. *Dairy Science & Technology* **2013**, *93*, 415–430.

Cleaver, J.W., Yates, B. The effect of re-entrainment on particle deposition. *Chemical Engineering Science* **1976**, *31*, 147–151.

Faldt, P., Bergenstahl, B., Carlsson, G. The surface coverage of fat on food powders analysed by ESCA (electron spectroscopy for chemical analysis). *Food Structure* **1993**, *12*, 225–234.

Fieg, G., Wozny, G., Buick, K., Jeromin, L. Estimation of the drying rate and moisture profiles in an industrial spray dryer by means of experimental investigations and a simulation study. *Chemical Engineering Technology* **1994**, *17*, 235–241.

Fletcher, D.F., Guo, B., Harvie, D.J.E., Langrish, T.A.G., Nijdam, J.J., Williams, J. What is important in the simulation of spray dryer performance and how do current CFD models perform? *Applied Mathematical Modelling* **2006**, *30*, 1281–1292.

Fletcher, D.F., Langrish, T.A.G. Scale-adaptive simulation (SAS) modelling of a pilot-scale spray dryer. *Chemical Engineering Research and Design* **2009**, *87*, 1371–1378.

Francia, V., Martin, L., Bayly, A.E., Simmons, M.J.H. Deposition and wear of deposits in swirl spray driers: The equilibrium exchange rate and the wall-borne residence time. *Procedia Engineering* **2015**, *102*, 831–840.

Fu, N., Woo, M.W., Lin, S.X.Q., Chen, X.D. Applicability of the reaction engineering approach for droplets of different sizes. *Chemical Engineering Science* **2011**, *66*, 1738–1747.

Fu, N., Woo, M.W., Selomulya, C., Chen, X.D. Shrinkage behavior of skim milk droplets during air drying. *Journal of Food Engineering* **2014**, *116*, 37–44.

Gabites, J.R., Abrahamson, J., Winchester, J.A. Air flow patterns in an industrial milk powder spray dryer. *Chemical Engineering Research and Design* **2010**, *88*, 899–910.

George, O.A., Chen, X.D., Xiao, J., Woo, M.W., Che, L.M. An effective rate approach to modeling single-stage spray drying. *AIChE Journal* **2015**, *61*(12), 4140–4151.

Guo, B., Langrish, T.A.G., Fletcher, D.F. Simulation of turbulent swirl flow in an axisymmetric sudden expansion. *AIAA Journal* **2001a**, *39*, 96–102.

Guo, B., Langrish, T.A.G., Fletcher, D.F. Numerical simulation of unsteady turbulent flow in axisymmetric sudden expansions. *Journal of Fluids Engineering* **2001b**, *123*, 574–587.

Guo, B., Langrish, T.A.G., Fletcher, D.F. Simulation of gas flow instability in a spray dryer. *Chemical Engineering Research and Design* **2003**, *81*, 631–638.

Handscomb, C.S., Kraft, M., Bayly, A.E. A new model for the drying of droplets containing suspended solids after shell formation. *Chemical Engineering Science* **2009**, *64*, 228–246.

Harvie, D.J.E., Langrish, T.A.G., Fletcher, D.F. A computational fluid dynamics study of a tall-form spray dryer. *Food Bioproduct and Processing* **2002**, *80*, 163–175.

Ho, C.A., Sommerfeld, M. Modelling of micro-particle agglomeration in turbulent flows. *Chemical Engineering Science* **2002**, *57*, 3073–3084.

Huang, L.X., Kumar, K., Mujumdar, A.S. Simulation of a spray dryer fitted with a rotary disk atomizer using a three-dimensional computational fluid dynamic model. *Drying Technology* **2004**, *22*, 1489–1515.

Jaskulski, M., Wawrzyniak, P., Zbicinski, I. CFD model of particle agglomeration in spray drying. *Drying Technology* **2015**, *33*, 1971–1980.

Jie, X., Chen, X.D. Multiscale modelling for surface composition of spray-dried two-component powders. *AIChE Journal* **2014**, *60*, 2416–2427.

Jie, X., Zhang, H., Wu, D., Chen, X.D. An improved calculation procedure on surface composition of spray-dried protein-sugar two-component systems. *Drying Technology* **2015**, *33*, 817–821.

Jin, Y., Chen, X.D. Numerical study of the spray drying process of different sized particles in an industrial-scale spray dryer. *Drying Technology* **2009**, *27*, 371–381.

Jin, Y., Chen, X.D. A fundamental model of particle deposition incorporated in CFD simulations of an industrial milk spray dryer. *Drying Technology* **2010**, *28*, 960–971.

Jongsma, F.J., Innings, F., Olsson, M., Carlsson, F. Large eddy simulation of unsteady turbulent flow in a semi-industrial size spray dryer. *Dairy Science and Technology* **2013**, *93*, 373–386.

Kaer, S.K., Rosendahl, L.A., Baxter, L.L. Towards a CFD-based mechanistic deposit formation model for straw-fired boilers. *Fuel* **2006**, *85*, 833–848.

Kar, S., Chen, X.D. Effect of high mass fluxes on heat had mass transfer through a flat surface. *Journal of Process Mechanical Engineering* **2004**, *218*, 213–220.

Kieviet, F.G., Kerkhof, P.J.A.M. Air flow, temperature and humidity pattern in a co current spray dryer: Modelling and measurements. *Drying Technology* **1997**, *15*, 1763–1773.

Kota, K., Langrish, T.A.G. Fluxes and patterns of wall deposits for skim milk in a pilot-scale spray dryer. *Drying Technology* **2006**, *24*, 993–1001.

Langrish, T.A.G., Keey, R.B., Hutchinson, C.A. Flow visualization in a spray dryer fitted with a vane-wheel atomizer: Photography and prediction. *Chemical Engineering Research and Design* **1992**, *70*, 385–394.

Langrish, T.A.G., Kockel, T.K. The assessment of a characteristic drying curve for milk powder for use in computational fluid dynamics modelling. *Chemical Engineering Journal* **2001**, *84*, 69–74.

Langrish, T.A.G., Zbicinski, I. The effects of air inlet geometry and spray cone angle on the wall deposition rate in spray dryers. *Transaction of IChemE* **1994**, *72*(A), 420–430.

Lazar, W.E., Brown, A.H., Smith, G.S., Wong F.F., Lindquist F.E. Experimental production of tomato powder by spray drying. *Food Technology* **1956**, *3*, 129–134.

Lin, S.X.Q., Chen, X.D. A model for drying of an aqueous lactose droplet using the reaction engineering approach. *Drying Technology* **2006**, *24*, 1329–1334.

Lin, S.X.Q., Chen, X.D. The reaction engineering approach to modelling the cream and whey protein concentrate droplet drying. *Chemical Engineering and Processing: Process Intensification* **2007**, *46*, 437–443.

Lin, C.X., Phan, L. A numerical study of both internal and external two-phase flows of a rotating disk atomizer. *Drying Technology* **2013**, *31*, 605–613.

Liu, W.J., Wu, W.D., Selomulya, C., Chen, X.D. Spray drying of monodispersed microencapsulates: Implication of formulation and process parameters on microstructural properties and controlled release functionality. *Journal of Microencapsulation* **2012**, *29*, 677–684.

Mansouri, S., Suriya Hena, V., Woo, M.W. Narrow tube spray drying. *Drying Technology* **2016**, *34*(9), 1043–1051.

Masters, K. *Spray Drying Handbook.* **1979**. London: George Godwin Limited.

Meyer, C.J., Deglon, D.A. Particle collision modelling—Review. *Minerals Engineering* **2011**, *24*, 713–730.

Mujumdar, A.S. *Industrial Drying Handbook.* 4th edition. **2014**. Boca Raton, Florida: CRC Press.

Nijdam, J.J., Guo, B., Fletcher, D.F., Langrish, T.A.G. Challenges of simulating droplet coalescence within a spray. *Drying Technology* **2004**, *22*, 1463–1488.

Nijdam, J.J., Guo, B., Fletcher, D.F., Langrish, T.A.G. Validation of the Lagrangian approach for predicting turbulent dispersion and evaporation of droplets within a spray. *Drying Technology* **2006**, *24*, 1373–1379.

Oakley, D.E., Bahu, R.E. Computational modelling of spray dryers. *Computer and Chemical Engineering* **1993**, *17*, S493–S498.

Ozmen, L., Langrish, T.A.G. A study of the limitations to spray dryer outlet performance. *Drying Technology* **2003a**, *21*, 895–917.

Ozmen, L., Langrish, T.A.G. An experimental investigation of the wall deposition of milk powder in a pilot-scale spray dryer. *Drying Technology* **2003b**, *21*, 1253–1272.

Patankar, S.V. *Numerical Heat Transfer and Fluid Flow.* **1980**. Washington, D.C.: Hemisphere Publishing Corporation.

Patel, K., Chen, X.D. Surface-center temperature differences within milk droplets during convective drying and drying-based Biot number analysis. *AIChE Journal* **2008**, *54*, 3273–3290.

Patel, K., Chen, X.D., Jeantet, R., Schuck, P. One-dimensional simulation of co-current, dairy spray drying systems—Pros and cons. *Dairy Science Technology* **2010**, *90*, 181–210.

Patel, K., Chen, X.D., Lin, S.X.Q., Adhikari, B. A composite reaction engineering approach of drying of aqueous droplets containing sucrose, maltodextrin (DE6) and their mixtures. *AIChE Journal* **2009**, *55*, 217–231.

Renksizbulut, M., Nafziger, R., Li, X.G. A mass transfer correlation for droplet evaporation in high temperature flows. *Chemical Engineering Science* **1991**, *46*, 2351–2358.

Sadek, C., Schuck, P., Fallourd, Y., Pradeau, N., Le FLoch-Fouere, C., Jeantet, R. Drying of a single droplet to investigate process-structure-function relationships: A review. *Dairy Science & Technology* **2015**, *95*, 771–794.

Schiffter, H., Lee, G. Single-droplet evaporation kinetics and particle formation in an acoustic levitator. Part 2: Drying kinetics and particle formation from microdroplets of aqueous mannitol, trehalose, or catalase. *Journal of Pharmaceutical Science* **2007a**, *96*(9), 2284–2295.

Schiffter, H., Lee, G. Single-droplet evaporation kinetics and particle formation in an acoustic levitator. Part 1: Evaporation of water microdroplets assessed using boundary-layer and acoustic levitation theories. *Journal of Pharmaceutical Science* **2007b**, *96*(9), 2274–2283.

Schwarz, J., Smolik. J. Mass transfer from a drop—1. Experimental study and comparison with existing correlations. *International Journal of Heat and Mass Transfer* **1994**, *37*, 2139–2143.

Southwell, D.B., Langrish, T.A.G. Observations of flow patterns in a spray dryer. *Drying Technology* **2000**, *18*, 661–685.

Southwell, D.B., Langrish, T.A.G. The effect of swirl on flow stability in spray dryers. *Chemical Engineering Research and Design* **2001**, *79*, 222–234.

Southwell, D.B., Langrish, T.A.G., Fletcher, D.F. Process intensification in spray dryers by turbulence enhancement. *Chemical Engineering Research and Design* **1999**, *77*, 189–205.

Ullum, T. Simulation of a spray dryer with rotary atomizer: The appearance of vortex breakdown. *Proceedings of the 15th International Drying Symposium* **2006**, pp. 251–257.

Verdumen, R.E.M., Menn, P., Ritzert, J., Blei, S., Nhumaio, G.C.S., Sorenson, T.S., Gunsing, M. et al. Simulation of agglomeration in spray drying installations: The EDECAD project. *Drying Technology* **2004**, *22*, 1403–1461.

Versteeg, H.K., Malalsekera, W. *An Introduction to Computational Fluid Dynamics: The Finite Volume Method.* **2007**. Pearson Education, New York.

Vigh, S.N., Sheehan, M.E., Schneider, P.A. Isothermal drying of non-nucleated sugar syrup films. *Journal of Food Engineering* **2008**, *88*, 450–455.

Wang, S., Langrish, T., Adhikari, B. A multicomponent distributed parameter model for spray drying: Model development and validation with experiments. *Drying Technology* **2013**, *321*, 1513–1524.

Wawrzyniak, P., Podyma, M., Zbicinski, I., Bartczak, Z., Rabaeva, J. Modeling of air flow in an industrial counter-current spray drying tower. *Drying Technology* **2012**, *30*, 217–224.

Woo, M.W., Daud, W.R.W., Mujumdar, A.S., Talib, M.Z.M., Wu, Z.H., Tasirin, S.M. Comparative study of drying models for CFD simulations of spray dryers. *Chemical Engineering Research and Design* **2008a**, *86*, 1038–1048.

Woo, M.W., Daud, W.R.W., Mujumdar, A.S., Wu, Z.H., Talib, M.Z.M., Tasirin, S.M. CFD evaluation of droplet drying models in a spray dryer fitted with a rotary atomizer. *Drying Technology* **2008b**, *26*, 1180–1198.

Woo, M.W., Daud, W.R.W., Mujumdar, A.S., Tasirin, S.M., Talib, M.Z.M. The role of rheology characteristics in amorphous food particle-wall collisions in spray drying. *Powder Technology* **2010**, *198*, 251–257.

Woo, M.W., Daud, W.R.W., Tasirin, S.M., Talib, M.Z.M. Effect of wall surface properties on the deposition problem at different drying rates in a spray dryer. *Drying Technology* **2008c**, *26*, 15–26.

Woo, M.W., Daud, W.R.W., Tasirin, S.M., Talib, M.Z.M. Condition of amorphous particles and deposits at different drying rates in a spray dryer. *Particuology* **2008d**, *6*, 265–270.

Woo, M.W., Daud, W.R.W., Tasirin, S.M., Talib, M.Z.M. Effect of wall surface properties at different drying kinetics on the deposition in spray drying. *Drying Technology* **2007a**, *26*, 15–26.

Woo, M.W., Daud, W.R.W., Tasirin, S.M., Talib, M.Z.M. Optimization of the spray drying operating parameters—A quick trial-and-error method. *Drying Technology* **2007b**, *25*, 1741–1747.

Woo, M.W., Daud, W.R.W., Mujumdar, A.S., Wu, Z.H., Talib, M.Z.M., Tasirin, S.M. Non-swirling steady and transient flow simulations in short-form spray dryers. *Chemical Product and Process Modeling* **2009a**, Article #4.

Woo, M.W., Daud, W.R.W., Talib, M.Z.M., Tasirin, S.M. Controlling food powder deposition in spray dryers at quasi-steady wall thermal condition: Wall surface energy manipulation as an alternative. *Journal of Food Engineering* **2009b**, *94*, 192–198.

Woo, M.W., Fu, N., Che, L.M., Chen, X.D. Evaporation of pure droplets in the convective regime under high mass flux. *Drying Technology* **2011a**, 29, 1628–1637.

Woo, M.W., Le, C.M., Daud, W.R.W., Mujumdar, A.S., Chen, X.D., Tasirin, S.M., Talib, M.Z.M. High swirling transient flows in spray dryer and consequent effect on modeling of particle deposition. *Chemical Engineering Research and Design* **2012**, *90*, 336–345.

Woo, M.W., Rogers, S., Lin, S.X.Q., Selomulya, C., Chen, X.D. Numerical probing of a low velocity concurrent pilot scale spray drying tower for mono-disperse particle production—Unusual characteristics and possible improvements. *Chemical Engineering and Processing* **2011b**, *50*, 417–427.

Wu, D., Lin, S.X.Q., Chen, X.D. Monodisperse droplet formation through a continuous jet break-up using glass nozzles operated with piezoelectric pulsation. *AIChE Journal* **2011**, *57*, 1386–1392.

Wu, D., Liu, W.J., Gengenbach, T., Woo, M.W., Selomulya, C., Chen, X.D., Weeks, M. Towards spray drying of high solids dairy liquid: Effects of feed solid content on particle structure and functionality. *Journal of Food Engineering* **2013**, *123*, 130–135.

Yang, S.F., Xiao, J., Woo, M.W., Chen, X.D. Three-dimensional numerical investigation of a mono-disperse droplet spray dryer: Validation aspects and multi-physics exploration. *Drying Technology* **2015**, *33*(6), 742–756.

You, X., Zhou, Z.H., Liao, Z.K., Che, L.M., Chen, X.D., Wu, D., Woo, M.W., Selomulya, C. Dairy milk particles made with a mono-disperse droplet spray dryer (MD2SD) investigated for the effect of fat. *Drying Technology* **2014**, *32*, 528–542.

Zbicinski, I., Li, X. Conditions for accurate CFD modelling of spray-drying process. *Drying Technology* **2006**, *24*, 1109–1114.

Zbicinski, I., Piatkowski, M. Continuous and discrete phase behaviour in counter current spray drying process. *Drying Technology* **2009**, *27*, 1353–1362.

Zbicinski, I., Strumillo, C., Delag, A. Drying kinetics and particle residence time in spray drying. *Drying Technology* **2002**, *20*, 1751–1768.

Zuo, J.Y., Paterson, A.H., Bronlund, J.E., Chatterjee, R. Using a particle-gun to measure initiation of stickiness of dairy powders. *International Dairy Journal* **2007**, *17*, 268–273.

Index

Printed and bound by CPI Group (UK) Ltd, Croydon, CR0 4YY

01/11/2024

01782617-0020